MANAGING INHERENT TECHNICAL ISOLATION (ITI)

A Strategic Approach to
Streamlining and Accelerating Leadership and
Technology Development within Organizations

DANIEL CHARLES

Copyright © 2022 Daniel Charles
All rights reserved
First Edition

PAGE PUBLISHING
Conneaut Lake, PA

First originally published by Page Publishing 2022

ISBN 978-1-6624-8119-2 (pbk)
ISBN 978-1-6624-8120-8 (digital)

Printed in the United States of America

To my wife, Jugbeh Charles, thank you for your love and
support. It is my hope that you will enjoy reading this one.
To my daughter, Paynetta Charles, your calming spirit and infinite
inspiration made this book possible. Please do not stop leading
the way to a better future. The world needs your inspiration!

CONTENTS

Foreword vii
Introduction xi

Chapter 1: Understanding Leadership in Technology
              Development 1
              The anatomy of a technology system 5
              System design 7
              Interface 9

Chapter 2: Understanding the Principles of Technical
              Leadership 15
              What is technical leadership? 15
              The critical elements of technical leadership 17
              The policy statement 26
              The plan's resolution 27
              Scope, objectives, and assumptions 28
              Damage valuation mechanism 30

Chapter 3: Project Management Competency and
              Technology Development 32
              Understanding and developing a project
              methodology 33
              Defining a project management methodology 34
              Defining and implementing key project
              management standards 37
              Project, program, and portfolio management
              standards 37

Governance tools and standards for technical project management 38
Project leadership and resources management 45
The role of measurement in improving project management capability 51
Project management office (PMO) capability development 54
Framework for implementing standard PMO 56

Chapter 4: Promoting Ethics in Leadership and Technology Management 74
Ethics in business management 75
Five critical steps to creating an ethical model 79
Ethical constraints and privacy challenges in the technology age 86
Incorporating ethical practices in technology development 90
Addressing the digital divide 100
The role of education in addressing the digital divide 102

Chapter 5: Enhancing Leadership through Efficient Technology Solutions Adoption 105
The positives and negatives of deploying ERP systems 106
Leading organizational ERP assessment project 109
ERP supply chain management solution review 111
Similarities and differences between SAP and Oracle SCM products 113
Evaluating ERP systems implementation success 117
Conducting enterprise ERP needs assessment 118
Evaluating ERP vendors and their products 121
Implementing enterprise ERP solutions 127

Acknowledgments 131
References 133

FOREWORD

Perhaps like many people, I love engaging conversations. What I find even more enjoyable are RAIN making conversations. Conversations that are impactful and, in special instances, game changers. RAIN is an acronym that stands for rapport, aspirations and afflictions, impact, and new reality (Schultz and Doerr 2011). RAIN was a sales methodology that I came across and began practicing somewhere between 2011 and 2012. As an expert in HCM (human capital management), with over twenty-five years of experience and helping thousands of companies with their employees, and secondly, an expert in Bible-based personal finances, with over twenty-eight years of experience helping thousands of families, and as an author of a best seller on personal finances, I have a perspective on people that is quite unique.

These two very different fields (HCM and Bible-based personal finances) are fields that, to most, would seemingly have nothing to do with one another, yet in my life, they are quite related and even inseparable. Simply put, I am an expert at studying and understanding people. After all, companies and businesses are made up of individual people. I understand what people think, feel, and what motivates them. This becomes a highly valued skill, especially when it comes to leadership development. I also have the ability to recognize when a "RAIN maker" comes along. Someone that is exceptionally gifted. That is what I see in Daniel Charles and the work he is involved with.

I met Daniel at my church about seven years ago. I was speaking at Mass on personal finances. After Mass, he approached me, introduced himself, and we had one of those "engaging" conversations.

DANIEL CHARLES

This casual new acquaintance quickly turned into a highly productive friendship that has lasted for almost close to a decade now. Our conversations went from engaging conversations to "RAIN-making" conversations. Daniel has quite the gift and passion for connecting with people (rapport), identifying people's desires (aspirations), zeroing in on the problem(s) that keep them from attaining their desires (afflictions), identifying the impact of the issues at hand, and creating a vision with that person as to how their problems and challenges could be met, and being able to solidly quantify that impact (the new reality).

Daniel and I have had our share of conversations about his profession (IT). A profession that, by any measurable standards, he is considered an expert. My expertise, at identifying problems within companies that are employee related, missed this one. I was astonished to have him identify, as a technical person, the inherent and major problem that exists today in IT. The problem of "inherent technical isolation" (ITI). Because of the fact that IT professionals are technical in nature, they tend to either be segregated from the companies and businesses they serve by leadership, or worse, these IT employees, managers, and leaders choose to isolate themselves in the companies they work for and with. Neither of these are beneficial nor productive. Through this isolation, IT professionals and company leadership are missing out on the true importance IT can bring to the overall success of the companies or businesses they serve.

The fact that a technical expert would not only be able to identify such a problem but would also be able to provide the solution to address it speaks volumes of the character he brings to his profession. IT professionals, IT leadership, and the leadership team in any sized organization ought to pay close attention to this because this is a RAIN-maker conversation (message). Daniel has identified and created a manual that addresses a major problem that plagues many, if not most companies and businesses. To understand what inherent technical isolation does, all one has to recognize is what is occurring in and to the business landscape today under the current conditions of a worldwide pandemic.

MANAGING INHERENT TECHNICAL ISOLATION (ITI)

Companies are struggling to adjust their business models to this "new" way of having to conduct business. The rank-and-file employee, as well as leadership at all levels, are also struggling to adjust to these realities. Isolation within any organization can affect the sales of its products and services, the marketing of those products and services, the transportation and logistics behind the delivery of the products they sell, as well as the service and support needed for those same products and services. The very bottom line of company profits can be made or broken by this issue.

Technology has the ability to bridge the gap and solve these important challenges of isolation, and IT is the driver behind this technological solution. It could be seen by many as ironic that the IT profession has the ability to bring the answer to the single biggest challenge facing business in our lifetime, but that answer is the same problem that has plagued IT since its beginning, the problem of inherent technical isolation.

To properly address this, IT professionals and business leaders must step out and up to embrace the call that Daniel's book proposes. In his book *Managing Inherent Technical Isolation (ITI): A Strategic Approach to Streamlining and Accelerating Leadership and Technology Development within Organizations*, Daniel challenges the reader to become more like *servant leaders*. A servant leader is a true leader. A true leader is one who looks to *serve* the people he leads while always demonstrating excellent virtue, morals, character, humility, and professionalism.

This book, in my opinion, will become a mainstay resource for IT and leadership practices, a strategic guide to solving a complex yet dynamic issue that is so prevalent in companies and businesses today. Not only should this book be read but also taken to heart and put to practice!

INTRODUCTION

Computers changed the way the world worked, providing automated processes that streamlined the organization of information. The trend toward managed information systems (MIS) became popular in the late 1950s as transistors replaced electronic tubes. By the early 1960s, large mainframe computers could be found within the cloistered walls of corporate America. These massive systems required specialized engineers to operate them. The main function of these computers in those early days was primarily for preparing financials and maintaining customer data. The media for storing information was often punched cards or large reels of magnetic tape. Due to the newness of automating reporting and managing data, organizations struggled to find a place to delegate the cost of these systems. For several decades following the 1950s, MIS typically was treated as a cost center, charging out the expense of operations to other business units. The MIS department was not given a place in the corporate management circle, most often being relegated to the accounting unit. The MIS manager typically reported to the chief financial officer. The computing leadership usually had no voice in the executive suite. This trend of managing computing resources continued into the 1980s as computers became user desktop systems capable of replacing mainframe functionality. However, this cost center without any representation at the executive level began to change as several innovations were introduced to technology. First, and possibly foremost, was the development of the Internet. The connection of various computing assets on a common platform, the World Wide Web, opened opportunities for companies to offer their services nearly anywhere. A subset of the Internet, social media, provided more in-depth access to the public as individuals from around the globe joined in. Organizations began to recognize the wealth of information web users were sharing within social media sites. Topics included personal information, likes and dislikes, hobbies, and, of great interest to businesses, the buying public's experiences with customer service and purchases. One of the best examples of this trend would be the feedback sections that accompany a webpage offering a product or service.

Companies quickly realized the information available on the Internet could enhance the internal data housed in various databases. The term *structured data*, internal sources of information, could be paired with comments and other records that Internet users shared, or *unstructured data*, to piece together a broader representation of an organization's customer profiles. In addition, a business could understand the public's sentiments or buying habits.

For decades, the computing environment within corporate America was generally treated as a provider of a service but had little or no presence in developing strategies or making decisions. The MIS department was most often an entity within another larger business unit. Data was processed and handed off for analysis. However, the relevance of the computing environment began to morph with the newly minted Internet. The terminology that had defined computing to that point was reinvented as *information technology* (IT) and began to find a path to the executive suite as new roles in the form of chief information officers redefined the activities of computing. Internal technology advances also played a part in this transformation with the introduction of networking, advancements in software development, and client-server applications.

The demand for information in support of the organization's core mission and strategies would introduce new concerns. One of those issues was the difference in skill sets that accompanied technologists. In short, those providing computing services were not of the same mindset as those working in the various other business units across the organization. The many information technology employees or third-party providers of computing services typically did not think in the terms of business success. Some of the apparent differences included

- lack of business background and knowledge,
- soft skills, such as communicating with others,
- interest in technology for its own sake, and
- personality differences in contrast to other skill sets in the organization.

This short list is typically the starting point for acknowledging the differences. Over the years, information technology employees were often expected to provide computing services while thought to offer no useful business ideas that advanced the company's mission or strategic planning. For information technologists, there was often no interest in understanding business concerns. Mission and strategy were relayed to the MIS department with the expectation they would deliver back data.

As an example, a traditional personality trait ascribed to software engineers/developers and computer operators has been the introvert. Introverts are most notably thinkers, often introspective, focusing on internal thought processes and ideas. Software developers exemplify the introversion as quiet, shy individuals writing code, hunched over a computer monitor. Another view suggests computer programmers are systematic and logical in processing thought. All these definitions tend to draw the conclusion that an introvert is a good choice to operate alone, in the quietness of the mind. There certainly would seem truth to the stereotype. However, something began to change this general perception in the late 1970s to early 1980s: the introduction of the personal computer in the home. Suddenly anyone, no matter their personality type, skill sets, age, or nearly any other characteristic or trait could "fiddle" with a computer. The most common activity in the early days was the development of code that controlled these wonders of electronics. Certainly, introverts were among the developers, but the availability of personal computers allowed others to join in. This hypothesis that anyone is capable of developing software was born out in a study by Lutes et al. (2009). The results of the study indicated there was no difference between the personality traits of extrovert versus introvert. Livingood (2003) indicated a similar trend when studying the personality traits of computer operators and network engineers. The development of computers that opened the world of technology to anyone also changed the scope of those who entered the career field of information technology professionally.

However, there is still a gap between an organization's technology providers and other business units. An anecdotal story brings to light the issue. As a banking organization began to expand data analytics into various business units, the information technology group struggled to understand the requirements of the reporting being requested. Many of the software developers in the organization had limited experience working directly with other staff. The soft skills necessary to interact with key stakeholders were lacking. However, in one case that stood out as a success, during the implementation of an automated system to process incoming loan payments, the information technology department learned lessons from prior requirements gathering. A software developer spent time at the desk of one of the loan payment processors. During the observation, several anomalies surfaced, including checks written in Spanish. The developer was able to communicate with the processing staff to better understand the many unusual issues that were a part of processing loan payments. The developer was able to go back to the IT team and add functionality that took into consideration various languages.

Despite efforts to better understand and mitigate differences in the characteristics of information professionals and the stakeholder community that relies on technology solutions, there remain perceived gaps that need to be resolved (Bassellier and Banbasat 2004, Freed 2014, Misra and Khurana 2017, Rasch and Tosi 1992). In recent years, university courses in information technology have begun to include talents such as "soft" (Schulz 2008) and other nontechnical skills (Danaher et al. 2019, Downing 2013). The divide or gap, whether intentional or not, can be termed as inherent technical isolation (ITI). This text describes the cause of ITI as the favored treatment of employees based on their technical or nontechnical skills.

MANAGING INHERENT TECHNICAL ISOLATION (ITI)

The idea of isolation within the workforce has been studied in the past. Farrar (2019, February 15), writing for *Forbes* magazine, discussed the behavioral health concerns of virtual workers. Of interest, the article appeared just one year prior to the outbreak of the coronavirus pandemic of 2020–2021. Although the piece focuses on remote working as the isolation factor, the behavioral issues are similar to those faced by technical employees that find themselves insulated from the rest of the business. Research studies also emphasized the depth of concern for *professional isolation* (Golden et al. 2008, Marshall et al. 2007). One of the earliest studies in technology isolation (Shuter 1984) suggested such workers were often operating alone. Shuter did remark that the technologist provided key support for business operations but was typically working in solitary. Studies have also been attentive to perceptions of the value of technology use, often a factor in the segregation of technology professionals and other business units, toward success in achieving strategic organizational goals (Lewis et al. 2003).

A more important concern related to overcoming the isolation factor for technology has been the more recent inclusion of representation of technical expertise in the executive suite. The position of chief information officer (CIO) began to surface in the early 2000s. Gendron et al. (2009) suggested the CIO provides a broader emphasis on integrating technology services toward meeting overall business strategic goals. By the mid-2010s, most corporations employed a CIO. The extension of C-suite information technology representation continued as the roles of chief technology officer (CTO), chief information security officer (CISO), and other titles appeared.

Despite the studies and articles noted, there has been little done in developing a framework to mitigate the ongoing isolation inherently lived by technology professionals. The concern for the lack of nontechnical skills among information technology professionals has focused attention on the need to develop a model for addressing the concerns. This text introduces a proposed framework for developing strategies to help reduce the inherent technical isolation (ITI) problem. The hope is that the information shared in this book provides a basis for organizational leadership to follow in implementing policies that work toward eliminating isolation of technical professionals from their nontechnical counterparts, especially in the execution of technology solutions.

CHAPTER 1

Understanding Leadership in Technology Development

For those who place a strong emphasis on leadership, it usually comes as a surprise when the outcome of a technology deployment or evaluation process within organizations frequently treats the role of effective leadership as an afterthought. Without any doubt, leadership plays a critical role in organization management, especially when it comes to developing, adopting, managing, and maintaining technology solutions. However, in today's business environment, there exists a culture or management practice that actively seeks to downplay the value of effective leadership in the implementation of technology processes. With regard to this culture, my experiences gathered from working with both Fortune 100 and Fortune 500 companies are very similar. On several occasions during my corporate career, whenever I asked people, especially nontechnical employees, about what role they played or would play in their company's development or adoption of new software solutions, the answers often follow a general theme which is "leave that to the information technology (IT) department. They will tell us what to do or perhaps do it themselves."

Many of the answers were a direct outcome of an existing business culture—a culture that I describe as inherent technical isolation (ITI). Today, ITI is sadly prevalent across many organizations. An interesting thing to note is that even collaborating team members, especially those with nontechnical backgrounds working on tech projects, cannot effectively explain the value link between the IT department, a technology solution, and their organization's strategic vision. For many, the technical complexity that often characterizes the technology process coupled with the unwillingness of most IT departments to effectively break down technical language are some of the reasons why people would prefer to let things be the way they are.

Nevertheless, the role of leadership in enhancing sustainable technology development is very important. To examine this importance, a deeper look at the value of leadership is critical. Across many generations, our pursuits to fully understand and assess the true values of human progress have led humankind to consistently seek and effectively examine the integral objectives of practical leadership with regard to individual and organizational needs. In many places, while professional development experts have always had conflicting submissions about the true meaning of practical leadership, one of the most recurring themes of collective agreement across ancient and modern times rests on the idea that leadership plays a critical role in the structure, composition, and implementation of development solutions, including software and other technical inventions. For example, in ancient Egypt, the Pharaoh led critical initiatives on this recurring theme. In Rome, consuls and senate members consistently constructed leadership models to support development initiatives using this recurring theme. And with each moment in history, the impact of effective leadership on technology advancement and maintenance cannot be overemphasized.

Today, with continuous advancement in technology, a major gap left to fill in the business and technology environment is the understanding of how technology leaders can effectively break down technical communication to help support enterprise-wide ownership of technology growth and development processes. In 2020, Gartner Inc. forecasted that worldwide IT spending will hit at least $4 trillion by quarter four of 2021. The new increased spending which is projected to be 8.6 percent up from the previous year is set to be primarily driven by rich and industrialized nations like the United States,

MANAGING INHERENT TECHNICAL ISOLATION (ITI)

Japan, China, the United Kingdom, Germany, France, and more.[1] This pattern of consistent increase in global spending on technology can also be broken down and effectively analyzed across industries to help determine the true importance of how and what areas organizations are looking to drive efficient technological changes. Given the nature of fast-growing investment in technology, organizations that are looking to compete globally would have to begin building new management processes that carefully integrate leadership and technology. This new management approach must endeavor to deliver a cultural shift where nontechnical leaders across organizations are positioned to play key roles in driving and supporting organizations' and businesses' technology needs. This book helps to break down the fundamental dynamics behind such cultural shift.

Accordingly, when organizations are seeking to deploy a new technology solution, it is essential for individuals and organizations to first understand the structure and composition of any desired technology. This is very important because many of the greatest failed technological inventions in history were largely due to inefficiencies relating to ineffective systems' construct as opposed to the belief that it was a failure of inadequate technology management practices. Today, for example, mobile, cloud, and virtualized workstation are part of the leading technologies pursued by several competitive organizations. Therefore, possessing a vital understanding of the construct and business values of these technologies is a fundamental leadership requirement necessary for adopting any needed business technology solution.

To begin breaking down the cultural shift toward effective integration of leadership and technology, an important historical foundation is needed. Throughout the history of technology developments, a dual approach to exploring and utilizing the value of tech applications has been increasingly critical. One school of thought, at least those who might be considered as relatively non-tech-savvy, continue to view technology consciously or subconsciously as a means to an

---

[1] Gartner Inc., "Technology Spending Enters a New Build Budget Phase," July 14, 2021.

end or perhaps a solution that is basically focused on the primary application of tools to achieve desired results. On the other hand, the more tech-savvy proponents do emphatically view technology's driving force as a function of human taste—a concept that is powered by our intellectual, moral, and business-specific needs, as well as our demands and values. Put simply, the latter school of thought fundamentally visualizes technology as a unique element of human construct, necessitated by an intelligent desire to support continuous progress across human development. Understanding these schools of thought is incredibly vital to one's intellectual development, especially regarding his/her use of, valuation, and appreciation of technology. Presumably, a logical description of the anatomy of a technology application is essential to support one's true understanding and appreciation of a technology solution.

To this end, this section of the book breaks down the concept of a technology system's anatomy. By reading and understanding the details associated with a software development, I believe that many people including nontechnical leaders would be able to clearly visualize the technology development process and also be able to make relevant decisions regarding the need for and management of a particular technology solution. The ultimate goal is to ensure that effective leadership drives the technology selection, deployment, and management processes.

The anatomy of most tech systems encompasses an *idea*, a *system design* (made of inner and outer layers), and a complete *user interface*.[2] As a quick point of information, the anatomy of a technology system is discussed substantially in several other high-tech books and articles. This book simply seeks to provide a primary understanding of the concept from a practical leadership viewpoint. The goal is to expand the leadership and technology integration process by providing key foundational understanding across both fields.

---

[2] William and Brown, *Computer Security Principles and Practice 2nd Edition* (Pearson, 2012).

The anatomy of a technology system

We begin with the *idea*. To understand this component, begin to visualize the concept of argument. The generic construct of an argument consists of an idea, analysis, evidence, and conclusion. The first step to developing an argument starts with identifying the proposition. For example, will Israel and Palestine ever achieve peace? This proposition, at least in its entirety, becomes the central idea around which the rest of the remaining constructs are developed. An application development also embodies such approach. While there are many essential elements applicable to the construct of a technology system, the application program notes would usually identify and define one or more of the below elements:

- *Program need.* The program need is a part of the application development process that requires research to justify both the historical and current conditions for which the new technology solutions are needed. From a general standpoint, it assesses the internal developmental competencies against the organization's challenges and summarizes the primary need or needs for the new system development. Similarly, from a business perspective, the program need evaluates current corporate capabilities and summarizes the critical need or needs required to support effective enterprise solution through a service approach adopted via the use of the new application. All of these are vital conceptual pieces that must be fully understood before any code is written. So do not simply think that a software starts directly with writing codes. On the contrary, it begins with why we need the code written.
- *Program objective.* The program objective is developed based on the results of the program needs analysis. Essentially, the program objective seeks to synthesize the program necessity across several business standards, including governance and strategic development processes. This also includes, but is not limited to, leadership competency enhancement,

systems improvement needs, process standardization, revenue generation, strategic portfolio management, and other critical business systems sustainability needs. Depending on the leadership focus, the most critical results areas are identified during the program objective, and objectives are drawn up to support the application development process. In short, the program objective further clarifies the program need and expands its supporting structure to help increase the potential value of the needed software.

- *Program value or commercial viability.* At this level, an assessment is carried out to fully examine the program need or identified program objectives. At the end of the assessment, critical business and developmental benefits are identified. A value assessment is then implemented to structure the benefits of the program need either in line with financial or nonfinancial value objectives. For example, some developmental organizations can pursue technology development simply as a legacy initiative intended to elevate corporate prestige, while others would primarily focus on market acquisition and long-term customer sustainability strategies. Whichever decision carries the emphasis, documentations are established to support the conclusions.
- *Program sustainability approach.* A clear foundation for the technology sustainability is examined at this level. The process begins with the program development. This approach is in fact identified through the sustained and systematic chain of events commencing from the program need up to its value characterization and subsequent acceptance of its strategic benefits. This approach, therefore, identifies structure, examines support, evaluates security standards, and assesses short- and long-term resource flexibility for the program continuation. While there are several other technical elements necessary to support the ideal construct relating to a technology system development, many of the underlying principles or critical outcomes relating to those elements would, in most cases, sum up a significant com-

ponent of the technical approach elaborated above. For example, some tech developers would focus the program note on a deeper analysis of the business case, which may include the organization's leadership capability, the institutional readiness with regard to current technology systems, the market analysis, customer value management, financial value management, and enterprise sustainability strategy. What this really means is that technology is not just technology but rather a business development culture that requires a collective buy-in from the rest of the organization. This should not be carried out via an ITI operation; we must collaborate to achieve this process together.

System design

Depending on the software development model selected (waterfall, iterative, rapid application development [RAD], etc.), it is important to understand, especially as a nontechnical person, that the project team or organizational culture must clearly inspire the model's selection. This is vital because some similarities of process implementation exist across the different models, especially a reoccurring theme that includes requirement analysis, unit development, and program testing. While these software engineering methods do specify straight system-level requirements, an emphasis must be placed on the leadership choice relating to system design or model selection. Accordingly, three strategic elements are executed during the technology system design:

- *Understanding requirements.* For example, a technology system's platform provides the capability for the collective operating environment and supports the effective running of the higher-level layer. However, the application of leadership requires that the system's critical requirements clearly address its purpose and proposed functions. Most technicians are not focused on a product's strategic value analysis; therefore, this must be resolved in advance as an important part of the leadership need required to support the technology development process.

- *Units development.* After completing the requirement analysis, developers implement programs called units. Each unit has a specific function required to support preestablished functionality resolutions. There are quality control measures, including coding specification, functionality requirements, output acceptance, and risk analysis, that must be done along each step. The leadership component at each level requires ongoing process evaluation to juxtapose outcomes against system requirements. Such an approach is needed to help avoid unwanted consequences. Put simply, the emphasis on leadership in the technical development processes does not seek to limit the value of the technician or the technical process itself. Rather, it seeks to demonstrate the critical benefit of utilizing effective leadership to assess and monitor implementation requirements against predefined standards.
- *Program testing.* It is important to note that computer program development starts with defining the purpose of the program. This requirement helps to validate the developer's language selection (Java, Visual Basic, Adobe Dreamweaver, Eclipse, etc.). The program-testing step provides valuable insights for assessing the system's user interface. However, while many technology users may view this process as a mere requirement for a product's development, the systematic software testing life cycle (STLC) process demonstrates an intensive process associated with this procedure. There are many books and articles on STLC framework. This book, however, only seeks to identify individual elements of the framework as well as the technical leadership processes needed to support the desired testing outcome. Like most methodologies, the six phases of the process include requirements analysis, test planning, test case development, test environment setup, test execution, and test reporting. To further expand one's knowledge on this subject matter,

this author recommends the text *Software Testing* by Gerald D. Everett and Raymond McLeod Jr.[3]

Interface

The third element of the system's anatomy is the operating system user interface, commonly known as the interactive application layer (IAL). Like the other elements above, the IAL is developed through a set of intense technical steps. Today, despite the growing calls for system standardization, most developers are often not creative, especially when the focus of their work is sorely placed on securing a bond with high-profile customers rather than truly meeting the ethical and leadership standards required to develop a good technology solution. Thus, the IAL is simply the component of the system that allows for meaningful interaction with end users. It consists of input and output functions that translate and break down communication to support enterprise productivity. In short, this is the component of the software that you work with to get a job done at your office or at your business.

Designing the IAL requires several technical undertakings, but as mentioned, this text seeks to emphasize an understanding of the leadership role, thus breaching the gap between technological complexities and business value analysis, especially as it relates to end users' utilization of software systems. If the reader wishes to further enhance his/her knowledge on operating system user interface development, this author recommends the text *The Complete Software Developer's Career Guide* by John Sonmez.[4] Meanwhile, in continuation of our quest to break down the leadership concept of a system's anatomy, below are examples of a few important leadership steps that can be applied to the IAL development process. By understanding the conceptual flow associated with these steps, even if you are a tech expert or not, you can effectively collaborate with your technology

---

[3] Gerald Everett and Raymond McLeod Jr., *Testing Across the Entire Software Development Life Cycle 1st Edition* (September 2017).

[4] John Sonmez, *The Complete Software Developer's Career Guide*, Lear Programming Languages Quickly.

teams to deliver value-added technology solutions to your organization. This is the goal of this book.

- *Define user context.* This level of the IAL development is tied to the program needs, which we have already discussed in the idea assessment section. When defining the context, the developer must bear in mind that the application's strategic purpose is to serve a greater business or leadership need(s) of the organization rather than just being another product to serve commercial needs. The IAL takes into consideration several factors, including future client information technology (IT) system capabilities, configuration needs, data migration, and other business functionalities such as the supply chain, accounting, and project management systems, respectively. In short, the application should exemplify all critical capabilities to effectively support the business needs for which it is being developed. IT and project leaders can further take control of this process through a detailed understanding of their service-level agreement (SLA). The SLA provides legal guidance that covers technical and managerial authority when implementing a project.
- *Define context requirements.* In developing the context requirements, a task list is developed to assess ways through which several critical functions, especially those that are going to be primarily supported by the application, are fully aligned and efficiently integrated with the organization's development process. Developing context requirements is key because most often, technicians only realize after the construct of an application that certain organizational systems are nonconfigurable. This is a big stage in the development process. A single mistake at this level can lead to multiple project setbacks, thus impacting other critical constraints such as time, budget, quality, and performance. Therefore, the final product from this level must be carefully evaluated by all the project leaders as well as the technical teams. Essentially, the proposed business systems

MANAGING INHERENT TECHNICAL ISOLATION (ITI)

that are critical to the application's support must be evaluated against all technical requirements. The final product should reveal both alignment and configuration flexibility, demonstrating that the future application has met the technical and leadership standards needed to fully implement the required business functions.

- *Assess requirements construct.* After assessing the functions and technical capabilities required to support the business needs, a design structure is put into place, reflecting a clear leadership and communication process that demonstrates the application functional capabilities. Before developing the structure, further detailed assessment is carried out across all proposed business systems, as well as the technology requirements. This assessment will look at input and output risks to the application, and will also further assess system-level capabilities, especially features that are required to support critical areas of the business or organization development. In short, the final product at this level should identify structure, risk, business and technology alignment, and a detailed communication process.
- *Develop and evaluate design structure.* This phase produces user interface designs, establishes clear interactive objects, and defines systems command applicable to the organizational processes or business needs and functionalities. Thought of more as the software engineering process, the entire framework of the IAL is built at this level. The strategic approach focuses on defining usability goals. Once those goals are established, the structural compositions are implemented in line with the predefined goals. For example, if the system is required to store both numeric and alphanumeric data with the greater objective of onward processing and evaluation, the design model will be constructed in accordance with a multifunctional layout. Sample data will then be used to evaluate if the storage and processing capabilities are fully enhanced to support real data application processing. Depending on the project's

timeline, new capabilities can be added, while others can be removed. Regardless of what is decided, the critical leadership required at this stage is to ensure that all predefined management standards as required for the technology are fully assessed against implemented activities. The design framework should then be evaluated for any further technical challenges, and once all critical requirements have been satisfied, the product can be moved for live testing or for further evaluations per the project management plan.

Before moving forward, let me once again emphasize how important the concept of effective technology and leadership integration is. To do this, I will share a quick story regarding my experience working in a supply chain role back in 2018. In the year mentioned, I worked at a metal distribution company that wanted to fully automate their production, inventory, and shipping processes. Prior to this process, however, employees across different departments consistently shared frustrations regarding the lack of information coordination and the prevailing conditions of ineffective leadership across the organization. Some of the challenges discussed included issues relating to how operational and managerial changes would be carried out, and weeks or perhaps months would pass before employees, especially those at the operational level, are informed about these changes.

Eventually, when the need for the automation process came about, the assistant operations manager was made the head of the project. His primary responsibility was to ensure that the project was planned and fully implemented to meet the overall objectives that senior management had developed. The overall impact of this project was particularly going to affect the way the processing, receiving, shipping, and warehouse departments conducted their daily tasks. As required, the operations manager had several high-level meetings with the central IT team and provided them with all the information he thought was necessary to support the project. Two weeks after implementing the application, employees from the various departments concerned were invited to a briefing/feedback meeting. As part of the meeting, the various team leads were instructed to give reports on the application usage.

As you may have predicted or not, the result of the application testing was a massive failure. The application functionality was, in so many ways, incompatible with the various teams' functions. In the shipping department, for example, the application was supposed to be able to scan steel certificates and execute an automatic data upload process to a central data location where shipping associates would then review the data and process shipping manifests as needed. While it is true that the application provided a means for scanning, it provided absolutely no means to verify scanned product to ensure that issues such as steel grades, cut lengths, original lengths, and many more were fully accurate. All these processes were vital to supporting customer information needs and value chain management. In the processing department, tags were wrongly labeled because many of the application inefficiencies I just discussed were seriously impacting effective operations, especially product processing and delivery. The result was also a chaos for the warehouse team since products were wrongly labeled and stored in locations that were thought to be right. In the end, the application was withdrawn and sent back for redevelopment. I honestly do not know how the process ended since I left the company a few weeks later to work on a new government contract at another company.

However, what is important to note is that these types of experiences are central to why I decided to write this book. Today, there are hundreds of companies spending millions of dollars on technology adoption with similar stories. In fact, some of the unwanted consequences from these failed technology processes have even led to individuals and organizations filing major lawsuits against some companies, while others have gone out of business due to persistent customers' dissatisfaction. Therefore, effectively managing ITI would also mean incorporating proven management practices that seek to clearly address management and operational inefficiencies, especially ones like the experience I just shared. These types of stories and many more provide the basis for the new business and cultural shift we are proposing in leadership and technology management. Essentially, if my formal employers at metal distribution had a clear leadership and technology integration framework in place, I believe that the investment would have been way less painful. Not only that but the application development process would have also been built from the bottom-up, thus utilizing the inclusive management culture as opposed to using the ITI approach. We need to make this cultural shift now!

Now that we have covered some of the key foundational elements relating to the technical construct of a technology system, let us move forward with some of the important leadership issues relating to technology management. Again, I want to emphasize that this text will not be particularly focused on all the technical details required for a computer system development. Instead, it will provide a synopsis of key technical steps while emphasizing the leadership and business management standards needed to support application development and to maintain its usefulness for the support of business needs. Accordingly, the next chapter focuses on providing a detailed understanding of the value of enhancing technical leadership to support organizational readiness for success, especially in regard to implementing technology projects. In the modern business environment, no organization or individual needs to be a victim of the poorly run ITI culture.

CHAPTER 2

Understanding the Principles of Technical Leadership

What is technical leadership?

Throughout the information age, history has proven that we often fail whenever we underestimate the role of leadership, communication, and collaboration in technology development or adoption process. While a business technology solution may be used to support several different needs, the technology strategic business purpose is to support constant innovation while maintaining accessibility, availability, and security of organizational data. Whatever we do, we must maintain this focus. From a management standpoint, this means that the role of effective leadership will continue to remain important in any technology adoption or development process. Put simply, we must stop promoting the backward inherent culture of technical isolation, a process via which companies consciously or subconsciously implement preferential treatment based on employees' technical versus nontechnical competency. Technology is a way of life. Today we are no longer just end users; we are partakers in the development and utilization of the world's technological innovations. However, I must also submit that to fully understand the reason behind this new cultural shift, a better understanding of technical leadership is highly critical. Therefore, we will begin by looking at technical leadership with regard to developing, managing, and maintaining technology solutions.

Technical leadership (TL) is the process of assessing baseline standards within organizations and combining the critical elements of innovation management, capacity development, and disaster management to support institutional competencies that promote lasting enterprise development. The process focuses on defining and integrating the capability of an organization's most critical business systems and assets to develop an internal independent strategic advantage. This strategic advantage is accomplished via a management practice that seeks to cultivate specialized functional capability within the organization. In short, technical leadership promotes critical development competition for the relevant organization actively pursuing this path. One of the most renowned examples of how technical leadership has delivered the value of independent strategic advantage across organizations is the case of the project management revolution.

Prior to the 1960s, business leaders implemented project tasks using formal approaches, yet with minimum or perhaps no structural framework that involved systematic startups, standardized data collection, and analysis. During the period also, there existed no record of a strategic integration framework that supported alignment of project objectives with overall corporate goals. However, this static or traditional functional management concept did serve the major business purposes as needed. Along those years, while relying on the strength of functional units' efficiency, organizational goals were set, requirements were defined, and deliverables were provided per contract terms.

Even though corporate leaders at that time believed in and greatly relied on such system, continual operational inefficiencies across the business environment occasionally sparked the need for a progressive transition toward a more reliable business system that integrated operational excellence with quality customer service management. Eventually, the project management innovation emerged as a critical field to support the missing link which was primarily summarized as the process of *strategic business objectives development, implementation, and management.* Sooner than later, the critical elements of project management, including defining project methodology, establishing baseline standards, developing metrics, formulating dashboards and scorecards, became crucial to meeting emerging business demands. Other challenges included standardizing project plans, managing global project portfolio, and many more. Going forward, these new constraints constituted some of the essential building blocks entrenched in the operational management principles of technical leadership. In succeeding chapters of this book, an extensive discussion is provided on technical leadership, covering detailed analysis of some of the essential elements, especially relating to technology development and project management. In this section, however, a brief introduction is given to provide the reader with an overview of the structural composition of the technical leadership concept.

The critical elements of technical leadership

Today, by assessing the outcomes of a number of development factors including public sector governance, private sector business systems development, as well as individual and collective competency development, we can be able to determine the extent to which technical leadership has impacted the growth and development of humankind. Put simply, one can even argue that the true power of technical leadership is still underutilized, especially given the promise of the world's technological revolution initiated in the early 1940s. Technical leadership does not only provide a path to get things done but also creates the ability to adopt critical innovative mechanisms that support the development of strategic business choices. Many of these choices are designed to promote and provide organizations with a new form of developmental approach which, even at its lowest, still provides competitive advantage.

Technical leadership looks at ways to enhance traditional leadership performance by assessing points of inefficiencies and adopting methods for technical improvements. Technical leadership consists of three central elements including innovation management, strategic capacity development, and business disaster management. Essentially, the effective integration and successful application of these elements is a critical task that can help to provide strategic transformation within organizations. While it is difficult to attain full technical leadership competency, the availability of multiple technology resources as well as the presence of business' information management system solutions often make achieving the highest standards of effective and efficient innovation management a reachable goal for businesses and organizations. The central elements of technical leadership are discussed below.

Innovation management. Mastering innovation management is key to achieving technical leadership success. Essentially, innovation management is a fundamental part of driving growth and development within an organization. Like most modern business concepts, it requires unique capabilities and careful considerations in both the design and implementation of strategies. Furthermore, given the ultimate importance of this field, companies that fail to adequately adapt and set new management standards to enable the effective application of innovation management are likely to face a series of critical business challenges. Moreover, since innovation management affects almost all the significant aspects of business-related processes, the need for proper innovation management cannot be overemphasized. Today, it is practical to suggest that companies with a robust innovation management strategy are highly likely to increase revenue, raise their stock value, and create more jobs as opposed to companies without one.

One of the most critical examples of innovation management empowered by the use of standard technical leadership is the systems application and data processing (SAP) innovation management system. As the global digital revolution continues to provide organizations and individuals the chance to combine information and technological resources, leading change across the business world requires mastering and effectively managing technologies like SAP, especially for competitive global organizations. SAP innovation management represents one of the breakthrough platforms in the field of digital revolution. From an operational standpoint, the application provides information solution through collecting, structuring, and evaluating ideas. SAP innovation management uses a combined strategy of coaching and flexible evaluation criteria to help individuals and organizations transform ideas into invention and subsequently transform invention into innovation.[5]

In a more specific way, SAP innovation management uses a set of organized procedures that can be utilized to propel cutting-edge business and leadership changes in the field of innovation management. The first component of the model or the flexible evaluation criteria can be applied to achieve strategic business objectives as follows. First, by applying key organizational leadership standards, innovation management helps accelerate ideas by turning them into understandable concepts. Next, the framework is institutionalized through creating and running an innovation office within the business which helps to improve organizations' ability to adapt, evaluate, and manage ideas throughout the development stage of a desired technology solution. When effectively cultivated, the innovation management office (IMO) can be used to address innovation gap across different functions of the business while fostering interdisciplinary team building for the development of better ideas. Also, SAP innovation management encourages speed and transparency in the innovation management and development process through constantly engaging employees with a focus on integrating clear leadership objectives with overall product development or service execution.

The second component of the SAP model, the coaching strategy, is highly useful to technical leadership because it builds on the principles of communication and collaboration in problem identification and delivering solutions. The SAP coaching model plays an important role in helping organizations overcome resistance to change through building a culture of communication excellence. Essentially, the model is applied in three specific ways. First, the model helps overcome resistance to change by providing speed and transparency along with constant employee engagement. This approach is highly needed in organization development since the

---

[5] Platinum DB Consulting, "The True Benefits and Strategic Business Values of SAP Innovation Management Solution."

need for employees' inclusion and transparency remains paramount to organization growth and development. The model does not isolate based on technical or nontechnical competency. Instead, it seeks to build collective competency by cultivating enterprise-wide technology, leadership, and management skills. Next, it examines the integral organizational culture of resistance to change and integrates resistance management mechanisms along the process of innovation management and project implementation. This step is ideal because it helps change managers anticipate resistance to change and prepare resistance management measures in advance. Finally, SAP innovation management coaching model is beneficial in overcoming resistance to change because it deals with resistance to change in a highly technical way. Consequently, through the coaching model, technical leadership questions such as "What to change?" "What to change to?" and "How to lead acceptable change" are all key areas explored in the concept of innovation management. Moving forward, the second capability to be discussed under the technical leadership framework is strategic capacity development. Unlike innovation management that focuses on process design, strategic capacity development deals with issues like capacity threat identification, competency needs analysis, and solutions development.

Strategic capacity development. The second element of technical leadership is the concept of strategic capacity development (SCD). Today, despite the growing popularity of strategic planning, there exist several organizations without institutionalized strategic planning management systems. In fact, many project or enterprise leaders who become interested in integrating strategic planning knowledge and tools with their various business needs often struggle to establish a clear business case for such an endeavor. While many studies have proven that effective strategic planning does indeed improve organizational systems and can even lead to enterprise-wide productivity, some business leaders have argued that continuous enterprise resistance to strategic capacity development grows from issues like the lack of organizational maturity, conflicting traditions and ideologies, poor communication management, refusal to accept perceived benefits, and the fear of power transfer as a result of possible structural changes. The fear of power transfer is very often seen as the leading factor for resisting change within organizations. Accordingly, the decision to adopt management standards on strategic practices has to be implemented through the use of intelligent business principles and effective leadership within the organization. Among other key business development needs, SCD helps organizations achieve four critical objectives including market advantage, improved efficiency, customer value management, and revenue generation. For example, two of today's leading organizations, Google and Salesforce, have greatly benefited from applying the SCD development approach. In the succeeding paragraphs, I will discuss how these organizations are applying the SCD concept.

Google, in particular, has benefited from SCD in many different ways. However, the company's strategic vision is often truly appreciated when looked at from a historical perspective. Google, a world leader in cloud-based services, was founded in 1998 by PhD students Larry Page and Sergey Brin.[6] The mission of the company is to organize world information and make it both useful and universally acceptable. Starting from a humble beginning, Larry and Sergey began Google as a research project in a small garage. Today, the company has a global luxurious headquarters located in Mountain View, California, and also owns multiple infrastructural assets around the globe worth billions of dollars. Google makes its revenue by combining and effectively utilizing the concepts of strategic business intelligence and operational excellence. From a strategic standpoint, Google assesses the potential of emerging competitors and, whenever possible, utilizes persuasive leadership mechanisms to purchase their businesses in advance. According to D'Onfro's 2015 article on Google's market competitiveness, since its founding in 1998, Google has purchased over 170 companies including YouTube, Motorola, Waze, and many more, spending billions of dollars and generating billions more in revenue.

---

[6] J. D'Onfro, "How Google is leading the world of acquisition through applying strategic business management innovation," Business Insider, January 2015.

Meanwhile, from an operational sense, the company makes money through advertising as well as offering multiple innovative business solutions to businesses all around the world. By using AdSense and AdWords, Google generates billions of dollars via these online advertising tools. In 2016, about 85 percent of Google's $68 billion revenue was generated from advertising only.[7] Google also generates revenue by offering several innovative business service solutions such as Gmail, Google Docs, G Suite, Google Maps, and many other independently developed applications which can now be fully integrated with Google's cloud services.

Moving forward on the SCD concept, Salesforce is another great example of how strategic thinking or the SCD's concept can quickly turn a regular organization into a global competitive force. From a humble beginning, the company used the strategic planning concept to cultivate technical resources in an effort to help deliver the sophisticated customer relationship management (CRM) platform it now shares with the world. In order to fully understand how Salesforce utilizes the SCD concept to drive success, let us first discuss the company's background and competitive business engagement strategies.

The company (Salesforce) operates in the technology industry with a key focus on customer relations management (CRM), providing software as a service (SaaS). In terms of financial standing, Salesforce was worth $155.08 billion and ranked number 285 on the 2018 Fortune 500 companies list.[8] The company also possesses a highly sophisticated network computing system that is strictly cloud-based. Some important aspects of the company's network infrastructure include the Salesforce platform which is made of important components such as data services and artificial intelligence systems. There are also robust platform technologies including Einstein's predictive intelligence and Lightning that are built and stored in a trusted multitenant cloud. Such a critical security approach offers

---

[7] J.,D'Onfro, "How Google is leading the world of acquisition through applying strategic business management innovation," Business Insider, January 2015.
[8] Fortune 500 2018 List of Industry Giants. "How Salesforce is leading innovation across the customer relations management market," *Forbes*.

a strong degree of data security for the company and, most importantly, its customers.

To better understand how Salesforce effectively applies the SCD concept is to review how carefully the organization aligns its operational activities with that of its central management structure for the purpose of supporting capacity assessment and competency determination and improvement. Through implementing this approach, the company offers varieties of customer relations management (CRM) solutions including Sales Cloud, Marketing Cloud, Service Cloud, Data Cloud, and Analytics Cloud, serving more than 250,000 customers around the world. These cloud-based solutions provide customers with access to various applications according to their business needs, and the computing data are managed and stored in the cloud. The company also has on-premise application solutions.

As indicated above, the company has a well-defined management structure and effectively practices clear separation of responsibilities. Each structure within the company, including the IT department, is set up to function according to straight guidelines that are in alignment with the overall corporate vision. In the implementation of business solutions, the need to integrate strategic business principles with customer value management is highly important. This is often seen when the organization's marketing section, for example, highlights the following capabilities as key to the company's marketing success: a systematic approach to revenue growth, customer-focused relationships, scalability, security, and mobile innovation management. The organization's management team also has clear rules governing the IT system and maintains critical security compliance certifications including CS Gold Mark, disaster recovery, and business continuity planning (BCP). These are standard leadership and management protocols that not only help to meet compliance but also improve internal capacity while driving excellence in service delivery. Other strategic business compliance and leadership standards include the Department of Defense Impact Level 1 and 4 security certifications, respectively. These institutional setups are vital to the continuous support of the business's long-term vision. In short, as a competitive organization, Salesforce remains committed to a vision built on the principles of effective technical leadership. When organizations master the science of SCD, it helps give them a level of competency that can be used to effectively cultivate the strategic advantage needed to become an industry leader.

Business disaster risk management. Across business operations, disaster risk management is a critical organizational goal for most competitive organizations. In fact, the concept is so critical to the point that most organizations without the necessary infrastructure as well as the technical human resource capability often result in outsourcing the responsibility. Nevertheless, because business risks are usually prevalent in the transfer and storing of data, it is important to build key internal understanding relating to the application of the business disaster risk management subject matter. To this end, this book addresses the basic principles of disaster risks management as a needed competency under the technical leadership framework. Accordingly, because this field is very extensive, no attempt would be made to teach the concept in detail. Instead, this book elaborates on why it is one of the three elements of technical leadership and how organizations can develop basic competency in this area. In the below paragraphs, a brief overview of business continuity and disaster risk management is discussed by looking at the concept from a practical implementation standpoint.

As organizations strive to push back costs, establishing business-critical functions and identifying the resources needed for service realization as well as conducting business impact analysis regarding corporate-related threats is highly critical. To this end, a business continuity disaster recovery (BCDR) plan can be highly valuable. Nevertheless, like any critical corporate endeavor, the cost of developing, testing, and maintaining an effective BCDR plan presents business leaders with a decision headache. For most organizations, it is never an easy task to effectively manage and allocate scarce resources across competing priorities. Put simply, BCDR planning is the management of a sustainable process that identifies the critical functions of an organization and develops strategies to continue these functions without interruption or to minimize the effects of an outage or loss of service provided by those functions when facing major business risks.

Disaster recovery is a technical component of the BCDR framework that focuses on business continuity management through leading comprehensive recovery efforts for data, telecommunications, and other technical infrastructure systems used within the organization. In a broader sense, a BCDR plan is a critical part of the emergency and risk management process of prevention, mitigation, preparation, response, and recovery plan developed by most countries' disaster management authorities and other relevant business institutions.[9]

When utilizing the technical leadership SCD's concept, proactive business continuity managers apply their energy beyond the production of the business continuity disaster recovery plan. They usually focus on a much bigger goal by seeking to transition beyond the traditional approach of having a static plan to running a comprehensive business continuity management system (BCMS). Such an approach ensures that the key elements of the emergency management system, including planning, preparedness, response, recovery, and mitigation, are fully functional and readily available. This approach also utilizes a strategic data management system to examine the organization's capability through looking at three major issues: (1) the current technology, (2) the available budget, and (3) the organization's actual capacity as it relates to meeting public expectations during a disaster crisis. A basic disaster risk management approach under the technical leadership concept would consist of four critical elements including policy statement, plan resolution, scope and assumptions, and damage valuation capabilities, respectively.[10] To further explain these elements, I will use situational examples, thus relating the implementation actions of a disaster recovery plan to a hypothetical organization. A brief discussion/examples of each of the disaster risks management elements are explained below.

---

[9] IBM, "How does a disaster recovery plan work? Develop a disaster recovery plan that boosts your cyber resilience and recovery capability."

[10] L3Harris Geospatial, "Disaster and Emergency Management Cycle. Developing critical management competency for disaster risk management."

The policy statement

The policy statement is highly effective when it is developed with the full backing of an organization's top management. As a critical component of the business continuity disaster recovery plan, the policy statement outlines the strategic framework that is used to lead disaster recovery implementation across the business. A good example of a disaster recovery policy statement is provided below.

> This plan focuses on cultivating wide experience from relevant industry knowledge sources. Our methodology takes into careful consideration the issues of organizational capability, information technology (IT) infrastructure and budget, and customers' expectations. These technical requirements are particularly important because they form the standard against which many BCDR plan are analyzed for the sole purpose of probing disaster recovery gaps. Most importantly, the success of this plan is tied to an effective capacity performance evaluation strategy as well as the implementation of critical exercises and testing requirements.[11]

A disaster recovery policy statement can incorporate several areas of need for the relevant organization. However, a focused policy statement should seek to address three key things. These include the strategic focus of the plan (what the organization intends to achieve), the internal capacity available to implement the plan (human and material resources), and the implementation control mechanisms (technical and managerial control measures that would be used to roll out and monitor implementation progress). As illustrated above, the

---

[11] Pretesh Biswas, "Example of Disaster Recovery Policy. A February 2020 article on how to plan your disaster recovery process with a focus on critical internal capabilities."

policy statement addresses four key components of the broader management framework, including defining the strategic picture (where the company hopes to go with the plan). Next, it identifies technology and human capabilities to be utilized in the implementation of the plan. Moving forward, it identifies operations and implementation gaps otherwise known as technical inefficiencies, and finally, it clearly defines practical exercises and testing as key evaluation and sustainability components of the plan. The next element is the plan's resolution.

The plan's resolution

The resolution is a technical management statement that communicates the specific purpose of the disaster management framework and identifies the technical approach through which an organization's mission-critical functions are identified and how such functions would be preserved and recovered in the case of a disaster incident. Below is an example of a generic disaster management plan resolution.

> A partial purpose of this plan is to support the business continuity management system. The business continuity management system outlines the overall coordination framework for disaster recovery management and control within the organization. In addition to the above, the complete resolution of this plan is to ensure that all the mission-critical functions of the organization as identified through the business impact analysis (BIA) process are given definite recovery time objectives (RTOs) using the needed resources, time, and strategies in the event of a disaster.[12]

---

[12] Pretesh Biswas, "Example of Disaster Recovery Policy. A February 2020 article on how to plan your disaster recovery process with a focus on critical internal capabilities."

Scope, objectives, and assumptions

The scope statement identifies the operational range and/or limitations of the disaster management plan. It clearly defines the specific setting as well as the requirements needed to formalize the implementation framework in preparation for a disaster management. One important point to note is that a disaster recovery plan can be implemented simultaneously across several different locations. However, the example outlined in this book is specifically related to a scope that is limited to a single state or a specific country. Moving forward, the objectives of the plan should identify specific operational steps that can be followed in order to achieve the plan's overall goals. Finally, the assumptions provide a set of probable conditions that could exist or must be in place in order to help support the disaster framework implementation. To further understand these elements, I have provided related examples below using a hypothetical case study. Here are some examples of a disaster management scope, objectives, and assumptions.

Scope. As indicated, the scope of this plan focuses on an organization functioning within a single state utilizing one location and operating a cloud-based IT infrastructure system. It should be mentioned that this plan does not incorporate any contingency for a disaster beyond a single location. In short, the plan's dimension under the BCMS is specifically focused on ensuring that there is an effective recovery team, management strategy, and operational framework in place at the specified location. Put simply, your scope should identify the plan's implementation location as well as the type of infrastructure available to support your desired disaster recovery goals.

Objectives. The objective of this plan is to carefully coordinate the recovery of all mission-critical functions within a specified timeframe, thereby ensuring that the business remains functional in the event of a disruption/disaster. This includes short- or long-term disasters or other disruptions, such as fires, floods, earthquakes, explosions, terrorism, tornadoes, extended power interruptions, hazardous chemical spills, and other natural or man-made disasters.

MANAGING INHERENT TECHNICAL ISOLATION (ITI)

Accordingly, the below specific objectives are examples of key priorities that can be used for a disaster recovery plan:

- Ensure the safety of all employees in the office building during a disaster.
- Mitigate disaster threats, or limit the damage that threats could cause.
- Have standard preparations to ensure that critical business functions can meet recovery time objectives (RTOs).
- Have documented plans and procedures to ensure efficient and effective execution of recovery strategies for all mission-critical functions.

There are several other examples that can be used. However, I have provided these few objectives only as a synopsis for informational purposes. Next, we will look at the assumptions of the plan, which is a very important component of the framework discussed in this book.

Assumptions. The need to implement a disaster recovery plan is often supported by a set of assumptions. The assumptions project possible scenarios that could help to facilitate the development of the disaster recovery plan. The assumptions are critical because they provide proactive steps and encourage implementation adjustments. In this book, I have provided a few of those options below. Thus, in regard to the hypothetical case being discussed, the viability of the business continuity plan (BCP) is based on the following assumptions:

- That a viable and tested IT disaster recovery plan exists and will be put into operation to restore data center service at a backup site within four to seven days.
- That the organization's disaster management department has identified available space for relocation of departments that can be occupied and used normally within three to five days in the case of an emergency.
- That this plan has been properly maintained and fully updated as required and that all members of the continuity management team understand the plan.

Damage valuation mechanism

The damage valuation step is the final strategic component of the business disaster risk management framework discussed in this book. This step is a very important subset of the technical leadership structure that organizations can develop and use to support disaster management. Today there are several models available to be utilized in the rapidly growing disaster management market; however, for this text, we will stick to two specific examples using the assumption that these examples are precisely unique to the potential disaster situation intended to be mitigated.

Assessing damage after a disaster is highly critical and often lays the foundation that could make or break the potential disaster recovery process. This task requires a deep competency. Some industry standards require that organizations have in place a viable framework for implementing such activity. Today most organizations often try to tailor their damage assessment strategies with the actual capability of the continuity management team. While this is a proactive approach, it must be noted that no matter the type of damage assessment employed, continuity management leaders must try to utilize an approach that seeks to balance speed and information quality. While there are several different assessment approaches that can be used, this book will focus on two of the most used techniques—modeling and comprehensive site assessment. Accordingly, the specific damage assessment strategies to be utilized under this plan are discussed below. Each of the examples provided below is only briefly discussed. Hence, there are two things to consider when reading the models: (1) the models are selected based on the specific case that needs to be mitigated, and (2) the models are chosen based on the continuity management team actual competency capability.

- *Modeling.* For this process, the continuity team first uses the predictive modeling approach to carry out an approximation of all damage conducted during a disruption (flood, earthquake, etc.). In the United States, as a means of ensuring information accuracy, progressive data collected via this approach are verified using industry best practices as provided by the Federal Emergency Management Agency (FEMA), the National Hurricane Center (NHC), and local disaster management authorities. The modeling assessment approach is important because it allows rapid identification of probable damages and provides a clear insight to review damage and develop cost estimation.

- *Comprehensive site assessment.* For this approach, the standards recommended by FEMA are also applied. However, in addition to the predictive modeling framework, a select group of highly trained disaster specialists often carry out a thorough site assessment at the disaster location. Authorized under a clear emergency protocol of the organization, the assessment team is tasked with the mandate of cross-referencing all data collected through the predictive modeling process. All data collected is then validated against predefined requirements, and the results are then used by the team for estimating the total cost of damage(s).

In summary, technical leadership is the process of assessing baseline standards within organizations and combining the critical capabilities of innovation management, capacity development, and disaster management to support institutional standards that promote enterprise-wide development. When applying technical leadership, corporate leaders look for ways to enhance traditional leadership performances by assessing points of inefficiencies and adopting methods for technical efficiency improvements. Once again, technical leadership combines the elements of innovation management, strategic capacity development, and business disaster risk management to adopt proficient management standards that support enterprise-wide productivity. This process also includes implementing continuous capacity assessment that helps create a long-term independent strategic business advantage.

CHAPTER 3

Project Management Competency and Technology Development

Business management is usually at its best when organizations are experiencing uninterrupted operational stability and financial success. Such was the case with a hypothetical US-based manufacturing company called Williams Machine & Tool for close to a century. A highly productive legacy system helped them dominate the market to the point of transitioning from a mere industry leader to becoming a pacesetter. The increasing level of success created a buoyant culture which, to an extent, became too difficult for competitors to ignore.

Essentially, the company was completely adamant that it had everything needed to meet both current and future market demands. They failed to incorporate any practical change management standards even when the competition was gradually changing around them. William Machine & Tool became the third-largest US-based machine tool company by 1990. However, the company's success was due to one product line of a standard manufacturing machine tool. The company spent most of its time developing ways to improve the single product line as opposed to pursuing diversified product opportunities. Even though some senior managers worked to identify this situation as a major weakness, several operational employees were particularly resistant to any formal change process.

As an outcome of the defiant management culture, the company failed to plan for any form of disruptive technological change. A few years later, a drastic drop in demand for the standard machine tool coupled with the drop-in revenue for tools (with or without modifications) led to the company being sold. Eventually, the clash of values between senior and middle-level managers resulted not only in the loss of jobs but also in the extermination of the company. The company's story and many others out there helped set the basis for continual project management improvement. Essentially, the ideal value of project management relative to this case would have been the assessment of current project capabilities, creation of standardized processes, and defining methods to improve competency across the entire organization in accordance with the competition. Effective project management by design functions within such scope and helps organizations build change management models to meet future competition demands.

Understanding and developing a project methodology

Corporate development is a continuous process. Management practices such as determination of project requirements, selection of objectives, and assessment of alternative measures aimed at supporting enterprise-wide development all form part of the critical tasks needed to support and ensure ongoing organizational development. In modern times, the gradual transition from the originally functional-focus business approach to the development of an enterprise-wide process-based approach vis-à-vis project management is truly a value-added initiative that supports effective business development culture. Pursuing excellence in project management is understandably an ambitious dream, especially given how difficult this can be. Nevertheless, the value it adds to individual projects as well as the entire organization can sometimes be immeasurable. In his 2018 article on project management innovation, Bridges puts it this way: effective project management reduces operational costs, delivers consistent results, increases competitive advantage, and limits inefficiencies. Nevertheless, none of the above outcomes are possible without a solid project management methodology.

The project management methodology, amidst other critical constraints, must establish the fundamental framework for planning, managing, implementing, monitoring, and controlling the project. Correspondingly, the official methodology must be structured in a highly innovative way to accommodate critical steps needed to deal with risks and other important project changes, especially including but not limited to managing what is widely known as the triple constraints (time, cost, and scope). However, beyond the effective management of the triple constraints is the fundamental challenge of risk. In the latter years after project management became widely recognized, one major phenomenon emerged. Essentially, it became clear that successful project managers had to first understand the context of the organization in relation to its strengths and weaknesses and be able to create a strategic framework for managing and dealing with risk at all levels of the project management. Thus, understanding and creating a clear project management methodology is critical to attaining project and enterprise-wide success.

Defining a project management methodology

Project methodology is a systematic framework that outlines the official process or processes via which a project can be implemented from start to finish. The concept of an effective project methodology is often characterized by key characteristics including but not limited to project planning and control, objectives clarification, analysis of alternatives, and constraints management. Basically, the methodology serves as the organizations' governing tool that is used to guide a project development and implementation.

Accordingly, a project methodology can be created by outlining clear implementation requirements across the following general project areas: (1) life-cycle phases, (2) the informal methodology or flexible project management approach, (3) standardized project tools, guidelines, and templates, (4) methods for capturing best practices, and (5) education and training plans and procedures (Project

Management Institute 2017). However, it is critical that the preliminary analysis leading up to the detailed project methodology includes an integrated leadership and management strategy that takes into consideration all risk factors, especially those identified via the business impact analysis (BIA) process. It is a well-known concept that an integrated project management strategy provides effective structural and technical competency to the project management process in a way that is sustainable. This entire process is governed and driven by predefined leadership standards approved by the organization.

The life cycle consists of a set of stages that is used by the project team to effectively manage a project. These stages include initiation, planning, execution, and closure. At the initiation stage, the focus is based on understanding critical project requirements, including defining goals, estimating deadlines, understanding risk, and determining project structure and labor requirements. The planning stage envisions and outlines most of the critical administrative standards needed to ensure that a project is implemented in line with an individual or organization's predefined project management guidelines. Once the key elements, including goals, deadlines, costs, and desired end results have been identified, the project then transitions to stage two.

The planning stage usually begins with the breaking down of large portfolios, programs, or projects into work packages and subsequently determining priority steps for future actions. The most essential task undertaken at this level is outlining responsibilities and timeline with the goal of setting into motion the project required activities. The project manager and his team are engaged with structural understanding of tasks and responsibilities. Cost, schedule, and scope clarification are carefully discussed at this level to ensure that the project is not only implemented as planned but that the intended value is fully derived from meeting the requirements as outlined in the project selection and justification. The decision-making process at this stage is typically informed by the business or project impact analysis, which will be discussed later on in this text.

Implementation of boundaries within a project management framework is developed based on knowledge gathered from the initiation and planning stages. Defining the project scope is critical to setting goals, developing objectives, hiring teams, determining critical success factors, measuring success, and monitoring performance. While the implementation stage is heavily focused on execution and performance management, many of the standards regarding project governance and control are also utilized at this level. Execution puts planned activities into motion, adhering to cost, time, and scope requirements. Portfolios and programs are broken down into project level to help provide functional clarity and process understanding. The project manager uses key control tools to help monitor performance while working to meet or exceed planned objectives.

The project closure stage is an internal review of completed or ongoing processes with the aim of successfully ending a project. Based on the team-preferred method, a desk review can be implemented, or a simple status update survey can be conducted across functional teams to verify project activities including cost, schedule, and collective performance progress. The entire project health is carefully accessed, and a formal status report is produced to determine the project reediness for closure. This report helps track effectiveness as well as the level of inefficiencies across the project management process. The project sponsor can also use the status update to determine what areas of the project should be closed immediately and other areas that should be reviewed, improved, and continued even after the current project is officially closed. Critical thinking is applied to the decision-making process to ensure that all conclusions regarding project closure or continuation are based on the business impact analysis which is linked to the overall corporate strategic goal.

The informal methodology or flexible project management approach is a strategy used to scan an organization's project management environment to determine what capabilities are available for project development, implementation, and management purposes. It guides the project implementation process by evaluating individual and collective competency level of technology, tools, and the system or framework available for project management.

Defining and implementing key project management standards

Project management is a complex field. The technical and administrative mechanisms needed to ensure the success of a project are not fixed. Basically, it always comes down to the combination of quality and experience coupled with the right tools as well as the proper management support. All of these attributes must be collectively and consistently applied over a period of time to derive a clear methodology for continuous project management success. Nevertheless, the project management standards adopted by any organization often require another vital piece which is quality leadership.

The role of leadership in assessing, deciding, developing, and improving management standards for business operations is critical in organizational development. Organizations that adopt project management standards often incorporate key practices to help facilitate project management activities across the business environment. Standards are important because they help develop a uniform culture across business operations. Ultimately, standards create accountability and provide a centralized management strategy through which operational activities can be implemented and measured. Across the project management field, there are several critical standards necessary for project oversight and control. Below we examine six of the most important leadership and project management standards that must be developed and consistently improved to help maintain a healthy leadership and project management environment.

Project, program, and portfolio management standards

Essential to understanding project management standards is the foundation of project, program, and portfolio management. Most project leaders and organizations sometimes consciously or subconsciously find it difficult to clearly separate and effectively apply these foundational principles. Therefore, the following analysis is provided to help articulate a basic understanding of these foundational elements. Project management focuses on the implementation of knowledge and the effective utilization of tools necessary to achieve the project requirements. Similarly, program management deals with the effective application of data, tools, and principles for the sole purpose of obtaining key benefits that would otherwise not be available by managing program mechanisms individually. Meanwhile, portfolio management provides an integrated framework for the management of one or more portfolios which includes projects, programs, and secondary portfolios, all of which are managed with the central aim of achieving strategic business objectives (Tucker 2015).

Each of the above elements contains several key characteristics that provide vital distinction rendering both technical and nontechnical management perspectives. For example, while project management focuses on meeting certain requirements such as start and end dates, personnel management, objectives realization, and more, program management, on the other hand, allows for the alignment of organizational projects and programs with the overall corporate strategic vision, supports risk management planning, and incorporates organizational change management strategies. In a unique way also, portfolio management is that key project management component which provides fundamental and critical guidance in the areas of organizational investment decision-making, project management transparency, strategic decision-making regarding areas such as but not limited to change management, resource allocation strategies, operational excellence, and customer value management.

Governance tools and standards for technical project management

Project selection. Organization leaders are often tasked with the responsibility of selecting a project. While this task may appear simple, failure to use critical and applicable selection criteria can often lead to multiple project management inefficiencies. There are many technical requirements for selecting a project; however, two requirements remain central and highly critical to understanding and building a framework for success across project management. The first one is the project cost analysis. A project should not be implemented void of the overall business vision. Basically, every organization's goal is to minimize cost and maximize revenue. To this end, all aspects of the business management including project selection and implementation must meet the requirement of cost-effectiveness and must also add value to the organization. The second major principle is the benefit-to-cost analysis. This concept often seems self-explanatory. However, when used in this context, it simply means that for a project to be considered ideal for selection and implementation, the potential value to be derived from the project must exceed the total cost of implementing the project. Regardless of the judgment standards used, if the project selection process does not take into consideration the two key criteria mentioned, it is highly likely that there would be unwanted consequences for the organization.

Project charter. Under the project management framework, the project charter exists as a management document that clearly defines the professional scope as well as the technical and managerial responsibilities of the project manager along with his team. Modern organizations have found a number of different ways to structure their project charter, and most organizations carry out this task in accordance with the business culture as well as the suitability of the project management office (PMO) in relation to the entire functional business system. Essentially, the project charter is the official document that provides a comprehensive strategic overview of the project scope, authority, and management structure. It also explains the project managers' relationship and operational scope to the entire organizations and to the various line managers in particular (Derenskaya 2018).

Project organization. Project organization provides the structure and/or framework via which the key activities of a project including but not limited to the project goals, objectives, schedule, and milestone are implemented within a particular organization or by a designated project management team. Through the project organization, businesses are able to define the structural framework upon which project planning, designing, implementation, monitoring, and control can be carried out. The project organization structure is the core tool that supports organization and management of all activities relating to the project. One of the most essential values of having an effective project organization structure is conflict minimization and resource maximization. According to Reynolds (2014), the core principle of project organization is to ensure the standardization of project management best practices, tools, and techniques, thereby creating improved conditions that can support the organization's strategic project management vision.

Project planning. Laudon and Laudon (2016) describe project planning as the management or organizational process through which a firm develops a set of predetermined course of actions necessary to support the achievement of a project's goal or the entity's overall corporate vision. Project planning can be as a result of external demand (contract won) or an internal pressure (need for improved management processes). As indicated, the overall objective of planning is to ensure the full attainment of a central goal. In project management, planning includes but is not limited to development of a project plan, establishing a project methodology, defining team members' qualification requirements, and developing a project budget. Planning also helps organizations to develop quality control processes that can be used to measure progress and to also develop corrective actions whenever objectives are not met. Essentially, good project planning can support the development of both strategic and operational strategies that can be used to effectively manage key project elements including project scope, project schedule, project cost, reporting metrics, and many more (Tesfaye 2017).

Project scheduling. Organizations are constantly looking for new ways to control project scope, present complex data to customers and policymakers, meet critical deadlines, and, most importantly, master the art of effective communication in project management. Nevertheless, managing the challenges that come with project management is not always easy. The project schedule represents one of the most critical elements in the field of project management. According to Leyman and Vanhoucke (2015), project scheduling is the application of tools, knowledge, and business skills to develop a project schedule. A project schedule communicates the tasks needed to be done, the part of the organization or persons responsible for the tasks, and the timeline for implementing the various tasks. Project managers use the project schedule to support an integrated approach to managing the timeline of a project. Basically, project managers can use the project schedule to allocate specific time to each component within the overall task list of a project while using a communications plan to help coordinate the effective implementation of each task. Essentially, effective scheduling is key because the project schedule represents time management. In short, without the project schedule, it is difficult, if not impossible, to control a project's length.

Project estimating. In project management, estimating can be a serious problem. Nevertheless, organizations that develop good estimating models often tend to manage cost well. Essentially, the complexities associated with estimating a project come in different ways. For example, within some organizations, limited professional experience is the biggest challenge, while for others, agreeing on the type of estimating model to use on a specific project can be a major problem. Project estimating involves the utilization of various applications to apply an estimated cost to each work package within a project. There are several types of project-estimating techniques. Some of the most popular techniques include analogy, parametric, and definitive estimating, respectively. Analogy estimation is based on comparison and similarity analysis. When using the analogous estimating technique, project managers analyze data from past projects that are similar to the current project. During the analysis, a detailed effort is made to identify the major differences between the two projects, especially in the areas of work complexity and technological change. After the analysis is completed, the cost is then adjusted to meet the realities of the current task. Parametric estimating, on the other hand, capitalizes on the use of statistical data. For example, if an organization won a contract to construct a mall in central Colorado, the project management team will make all effort to gather critical information relating to the costs of homes in the area and would also extend their analysis to gather information on material costs as well as transportation costs. All the data gathered will then be carefully analyzed and subsequently used to develop the current project estimates. The definitive technique uses tools like vendors' quotes, complete project plans, unit price analysis, market surveys, and consumer pricing reports to develop a comprehensive data package for the purposes of developing a complete estimate (Warburton and Cioffi 2016).

Monitoring and control. One of the most critical aspects of project management is the issue of monitoring and control. Usually, the foundation for successful monitoring and control is laid at the beginning of the project. Whenever project managers evaluate a project's statement of work (SOW), they try to seek sufficient clarity from the project sponsor/customer so as to adequately inform the development of smart goals and objectives. Smart goals and objectives are very important because they directly support a project monitoring and control system. According to Hazir (2019), project monitoring and control refers to the development of applications and tools to support the measurement and analysis of data with emphasis on predetermined project objectives against implemented tasks. The author argued that project monitoring and control is a critical foundation for project management success since it provides project managers with firsthand data relating to a project's performance.

In order to support a robust monitoring and control process, many organizations develop a comprehensive framework. To this end, one of the most popular frameworks available today is the management cost and control system (MCCS). The MCCS constitutes five major control stages including planning, work authorization and release, cost data collection, cost analysis, and customer and management reporting. The MCCS system integrates the three most critical project management elements including time, cost, and performance and provides a management platform to collect and report real-time data on each of these elements. Essentially, the MCCS does not only focus on cost control but also incorporates a monitoring system that provides data regarding a project's schedule and work performance. When project managers use the MCCS framework, they focus on an integrated approach to collecting, managing, and reporting data on cost, time, and performance (Dashkov and Tislenko 2018).

Project closure. Closing a project is not always an easy task. Nevertheless, when properly planned, project closure can be a smooth step in project management. Unfortunately, not all project managers possess the ability to adequately plan for successful project closure. In project management, gate review meetings are a form of project closure that can be used to either close a project's life-cycle phase or the entire project. There are key elements involved in the planning of project closure. First, the project manager and his team must plan for project closure by gathering, analyzing, and disseminating important timeline information to all stakeholders involved with the project. For example, all changes made to the project scope which may impact key deliverables dates as well as the project schedule as a whole must be properly communicated to all relevant stakeholders. This can be done with the use of forms, templates, and checklists. Next, the closure plan under a gate review style accounts for both contractual and administrative closures, respectively. Contractual closure takes place when the project manager along with the contract administrator verify and sign off on documents testifying that all deliverables have been met and that all action items have been fulfilled. The administrative closure, which comes after the contractual closure, involves the updating of contract records such as the closing of charge numbers and work orders. Essentially, effective project managers align the project closure plan with the project day-to-day activities in order to support effective closure at the end of the project (Choi 2018).

Communication management. Central to successful project management is the objective of maintaining effective communication across the project landscape at all times. Effective communication ensures that all relevant parties involved with a project receive the right information at the right time. To support effective communication management, the project manager and his team must develop a communications plan. Developing a communications plan is key because along the information channel, there exist several barriers to communication that can impede effective flow of project information from one person to another and/or from one group to the other. For example, some of the key barriers that affect effective communication include but are not limited to the sender's credibility, personal interest, interpersonal sensitivity, attitude, emotions, and self-interest. Though many of these barriers are personal, they can easily become institutionalized, especially when a project lacks a robust communication plan. As indicated, developing a communication plan involves the coordination of three key elements including communication policy, tools and technology, and people and processes. The communication policy outlines all the standard practices for information gathering, reporting, and dissemination. Communications tools and technology provide an avenue for information transmission from one person to another. Lastly, project personnel utilizes the communication policy, tools, and technology to follow processes and standards in order to support effective communication across project management. All these techniques are applied to support the central goal of achieving effective communication. Albeit, achieving effective communications management is a practice that comes with multiple experiences. Essentially, this is achieved by combining quality leadership and critical communication skill sets from a team of experienced project management professionals over time (Ruehl and Ingenhoff 2017).

Risk management. Unlike the past where most of the critical activities relating to risk were still unknown, today, most renowned project management systems are developed with the capability to effectively deal with risk. Risk represents a probability and a consequence that a project's goal might not be achieved (Kerzner 2017). For example, if a project manager sets up a target to introduce a new technology product within six months, the likelihood that the product will actually be introduced within nine months instead of six months could represent a risk facing that project. Accordingly, understanding what risk is creates the opportunity for project managers and organizations to collect important data for the purposes of developing a risk management strategy. Risk awareness often adds value to the organization in the sense that it creates a level of consciousness that is needed from both top management and the project team to support the development of a risk management strategy.

A risk management strategy is a framework designed to identify, analyze, monitor, and control the probability and consequence of a particular risk. Since risk is often a future probability, many organizations focus their risk management strategies on developing proactive measures that can be used to respond to a specific risk situation. For example, almost all human systems are constantly faced with major threats such as actions from malicious individuals, natural threats such as earthquakes, floods, etc. Essentially, the actual probability of a risk occurring is also based on the level of control put in place by the project manager or organization. As indicated, the major goal of the risk management system is to identify, analyze, and develop robust monitoring and control mechanisms that can be used to combat all forms of risk facing a project. In short, a comprehensive risk management system always supports the attainment of project objectives by effectively monitoring and controlling risk (Rozzani 2017).

Project leadership and resources management

Today, many project leaders are beginning to agree that ineffective leadership is perhaps the biggest risk to achieving project management success. Historically, a lot of thinking, as it relates to effective project management, was placed on the need to have standards, guidelines, processes, etc. as opposed to having the level of experienced and mature leadership that would set, assess, drive, evaluate, and improve project management capabilities. In short, an important layer of the project management environment was often always overlooked. As project management expanded, tracking the challenges of ineffective leadership was made possible by identifying the project environment. The project management environment consists of the policies, technology, and people (leadership). Without effective leadership, adopting the appropriate technology as well as implementing prudent standards to align with and support the organization's growth and development would be almost impossible. Therefore, the importance of an experienced and strategic leadership team in project management cannot be overemphasized.

For organizations trying to compete in the global market, a consistent management process must be developed to support the goal of attaining strategic advantage. Strategic advantage helps organizations to maintain operational excellence, improve service delivery, build better relationships with customers. It also improves product development processes and effectively manages risk. Nevertheless, all of these cannot easily be achieved if the organization cannot successfully control its costs. Unlike the pre-project management era, cost management today extends beyond monitoring and reporting cost data. It requires the act of verifying challenging data and adjusting major complexities to support cost-effective business decision-making processes. Essentially, given that resources are limited, effective competition management is indeed required to help support prudent cost management actions. For example, developing an enterprise mechanism to support estimating goals in project management in itself is not an absolute solution. To take this further, it is the assessment of the methodology to ensure that its application is fully generating the planned cost objectives needed to support efficient resource management that matters most. Project leadership and technical resources management is therefore a process that provides individuals and organizations with the ability to cultivate technical skills that help address inefficiencies across project management while using innovative leadership to control and effectively manage cost weaknesses.

As discussed, the role of leadership is to set the standard for continuous progress. A viable leadership will continue to innovate and develop quality management processes to support enterprise-wide strategic and operational development. In pursuing this path, several technical and administrative management standards can be derived. A few of these standards are discussed below.

Depending on the organization's structure, the group that is usually tasked with the responsibility of setting critical standards plays a critical role in laying the foundation for the project success. For example, project estimation is one of the most challenging areas in project management. This is so because when final estimates are wrong, it impacts the project negatively. Essentially, this could also affect not only the project manager's reputation but also the entire company. To this end, a high-level technical and managerial accuracy is needed whenever it comes to developing a project estimate. Inaccurate estimates represent critical risks to project management and can impede the overall scope of the project by directly affecting scheduling and cost. Hence, a robust and methodical approach that utilizes verifiable data including accurate past and current estimation techniques is highly advisable. While this is easier said than done, organizations have a responsibility to ensure that adequate due diligence is always applied to such a process. Some of the critical management standards that a leadership team can set for effective project estimation results are as follows.

Set estimation standards for project success. The project scope provides the fundamental requirements upon which other important project elements are developed. For example, the project charter, which is a critical initiating document, provides strategic analysis regarding a project's summary milestone schedule. Correspondingly, since work packages are developed based on the project scope, any wrong estimation, especially relating to a critical task, can certainly affect the project start/end date as well as the schedule and cost. In order words, all identified complexities should be carefully analyzed for the purposes of determining the actual extent to which the project can be impacted. The work breakdown structure (WBS) reveals the actual degree to which any risk can impact the project. The WBS covers the project schedule, cost, and scope. Moreover, by directly affecting a project's start/end dates, a wrong estimation could lead to new requirements, thereby directly impacting the project's scope. A critical relationship exists between scheduling, cost, project scope, and performance. Essentially, this dependent relationship exists because any direct effect on scheduling does not only impact cost but also affects the project's scope. When a negative risk occurs on any of these elements, it often leads to requiring fundamental changes not just at the WBS level but also at the overall project and program levels sometimes.

The impact of wrong estimate on project schedule. In project management, the technical and/or rational analysis supporting a conclusion that reveals the longest or shortest path to accomplishing project-related tasks within a specific schedule is referred to as the schedule's critical path (Haije 2017). The critical path shows the longest or shortest distance between the start and finish dates for a project by taking into consideration resources constraints, performance challenges, assumptions, work complexities, and more. For example, when a critical path is determined to be at least two weeks longer than the required task completion date, project managers assume that all conditions will remain constant. The assumption is vital because in the case of any change, the duration of the critical path may also change. Nevertheless, assuming things remain constant, project managers can utilize the critical path analysis to reallocate tasks and resources across the project for the purposes of achieving maximum resource optimization. To a very direct extent, wrong estimation can affect the schedule's critical path by simply impacting the task duration analysis. When project managers develop the critical path analysis, they validate task duration based on verifiable data. In the case of a wrong estimate, several different things could happen. First, the estimation team could seriously overlook the project's complexity; secondly, the project scope could be misinterpreted; and finally, the technology or resources required for successful project completion could be grossly understated. Therefore, it is important to ensure that the right steps are taken to avoid any future disaster relating to project estimating.

Project estimating should not always be considered a generic process. This is important because the estimation problem is usually not fixed. What individuals and organizations must understand is that the right project estimate should be pragmatic. This is so because the estimation technique is often highly linked to the relevant project's need or needs. To be able to adapt to such a flexible approach to effective project estimating, organizations or project teams should possess the capability to implement critical estimation assessment. Assessing deliverables to support a desired estimation choice is key. An assessment can be done in several different ways.

Assessing and utilizing the right estimating technique. To apply effective project estimating, a critical analysis of certain important background information including but not limited to professional reference materials, market and industry analysis, reference data on project knowledge and processes, etc. should be the first step taken by the project team or organization. Across different cases, it is important to note that the resultant project management culture will inform the eventual outcomes. For example, in certain organizations, the lack of adequate information relating to actual source documents needed to examine the foundational materials that can be used to develop final estimate for a project is often not available or technically incomplete. The unavailability of such documents can also lead to wrong estimation. For example, in one particular project management estimation case, the estimation team had claimed to have used the three-point estimation method to provide the project estimate. I will explain the three-point estimation method in a later paragraph. However, according to a review of the estimation document, the limited justification, especially as it relates to the accuracy of the final estimate, raised so many questions from the evaluation team.

First, the evaluation team reported that the triangular approach under the three-point estimation method which was used by the project team was completely flawed. To better understand this, let me explain what the three-point estimation method is. In project management, the three-point estimation process suggests that all estimation results including *most likely*, *pessimistic*, and *optimistic* scenarios are likely to occur if the assumptions and constraints provided at the start of the project remain the same. In this case study just discussed, a review of the file by the evaluation team concluded that the assumptions were significantly changing along the project implementation, yet the estimation team failed to consider the possible impact of the changing conditions.

Next, according to the evaluation document also, a review of past data associated with a similar project clearly reveals that the estimation team grossly overlooked the complexity of the current project and failed to adjust data for technology and tasks changes. In view of this analysis, it was clear that the failure to apply adequate due diligence, especially regarding the adjustments of work packages to properly account for technical and administrative complexities, easily resulted in the development of an inaccurate estimate. Therefore, it is strongly recommended that any estimation method to be used in any project management process be derived in accordance with the relevant project realities. This approach is carefully carried out based on a review of technical deliverables in reference to the official project scope. The process must take into consideration capacity constraints and must utilize the index cost verification method to adjust activities for technological and capacity changes with reference to a specific project's needs. In the end, a project team can benefit from this approach by correctly adjusting project requirements to account for critical tasks complexities as needed.

Pros and cons of project estimating process in organizations. Despite the importance of the estimation process, some organizations or project management teams prefer not to engage in any professional project estimating. One of the biggest arguments in support of this ineffective approach to project management is usually communicated as follows. Estimates are never exact; we will have to change scope as needed based on new project demands. Therefore, we must engage the project management process on a piecemeal basis and adjust cost resources as constraints change. This type of approach is usually relevant for projects that have not been done before. For example, the development of a new medical invention to cure an outbreak or perhaps a research led to inventing any breakthrough new technology that has never been done before. In the absence of these conditions, all projects must be effectively managed with the fullest understanding that resources are limited, and cost management is a fundamental requirement for growth and development. To this end, the project estimating process includes both pros and cons.

Project estimating is a technical process that requires the right structure, detailed planning, and adequate data. To support the overall project management objective, especially as it relates to developing accurate estimates, these three elements mentioned must function cohesively and effectively. Accordingly, some of the major pros of effective project estimating process in organizations include (1) an established structure, (2) a methodical process, (3) communication system/easy transfer of task from one station to another, and (4) experienced professionals with the capacity to perform identified requirements. However, in the midst of the positives, there also exist some important management inefficiencies including (1) ineffective top-down communication system, (2) a disintegrated estimation process which is mainly implemented as a functional unit task rather than a business's project management process, and (3) lack of a standardized estimating approach that cuts across an organization's project management environment. In order to capitalize on the strengths of the estimation process, it is highly recommended to develop cross-functional teams that can develop integrated project management estimating processes to support all sections of the organization. This process must be standardized and reviewed regularly in order to adjust techniques and incorporate new industry changes whenever necessary.

The role of measurement in improving project management capability

In project management, metrics play a critical role in ensuring that project standards are met by communicating the true status of the project. Selecting metrics is a delicate task due to the technical complexities involved with such a process. Accordingly, the value of metrics covers several important considerations in project management, especially when it comes to defining and evaluating requirements to be measured against key standards or baseline requirements. When we emphasize metrics, the goal is to draw attention to some of the key values metrics can provide in supporting effective project management capability. Metrics serve several critical purposes including assisting in determining the likelihood of success or failure, providing useful information to examine a project's status, verifying project health, supporting proactive project management, and detecting early risk signs for possible corrective actions.

While it is true that metrics can support critical business and project needs, it is also important to state that metrics are only effective if the communication and selection processes are carried out in a well-coordinated and truly strategic way. Basically, the role that communication plays in supporting project management success cannot be overemphasized. In using metrics, some project managers endeavor to provide a critical understanding of the value of metrics by seeking to validate the importance of project metrics in reference to achieving project management success. They build initiation strategies that detail out and examine metrics selection processes against specific project needs. For example, one of the repetitive and important areas in information technology management today is software upgrade. Understanding metrics selection for managing upgrade projects is a critical skill desired across information technology departments. An effective software upgrade project discusses the value and basic needs for both qualitative and quantitative project-based metrics based on the given project situation. The leadership process supports the need for incorporating stakeholders in the process of defining project requirements and documenting what constitutes project success. Explanations are given on how effective metrics can help to detect early warning signs and provide vital information to help support and improve clients' or stakeholders' relationships.

Use of metrics in a software upgrade project. Even with the existence of a fully functioning enterprise project management methodology (EPM), project management should still be implemented via a dynamic rather than static approach. Some authors have argued that the multiplicity of conditions that exists in project management makes it very difficult to assume that a single set of management tools or standards can be used as a one-size-fits-all mechanism at all times. In addition, the growing level of competing constraints in today's project management environment makes the process even more complicated. As such, an organization pursuing a software upgrade project can utilize the metrics management to systematically improve its project measurement capability. The goal of the metrics in this example below is to support the following critical project management standards including time, cost, quality, technology availability, performance, communication, and stakeholders' management. While these are highly recommended, as a project progresses and conditions change, adjustments should be made to either retire, adjust, or incorporate new metrics to meet real-time demands.

In a real-world approach, the process of selecting metrics for a software upgrade should basically focus on the critical success factors of the project in reference to the organization's strategic needs. The benefits of this approach are enormous because critical success factors often directly impact project success and can lead to immediate or potential scope validation based on the level of impact. When organizations define critical success factors at the start of a project, this process sets the basis for developing key metrics to support effective measurement. There are several areas that should be taken into consideration for a software upgrade project. Some of which include quality, communication, cost, leadership, stakeholders' relations management, and many more. To this end, some of the critical metrics that could be used for a software upgrade would include

1. number of milestones implemented according to schedule,
2. total cost per each requirement and estimated cost at completion,
3. efficient and effective use of resources in relation to planned objectives,
4. availability and durability of the necessary hardware to support the upgrade,
5. quality of personnel and overall project management performance,
6. level of effective communication, and
7. level of effective stakeholders' relations management.

Another key objective is to understand why these metrics are selected and what particular benefits they can provide the IT team or organization in helping to manage the project. Here are some of the important benefits to know. They include the following: (1) to support effective measurement of the project standards including time, cost, quality, technology availability, performance, communication, and stakeholders' management, (2) to effectively communicate with and achieve stakeholders' confidence in the project, and (3) to support both the business and project management objectives and to also help in the measurement of the project's critical success factors including time, cost, performance, and quality.

The importance of using quantitative and qualitative metrics. The metrics we have talked about consist of both qualitative and quantitative measurement standards. Within different aspects of the project management, each method provides different needs. For example, in most cases, stakeholders' communication or satisfaction is a form of standard that hugely requires qualitative measurement. This is so because it is the quality of response to project deliverables that matters most over quantity. When you have actually identified stakeholders' roles and needs relating to a project, your goal is to drive satisfaction through efficient and effective delivery of goods and services. At this point, you are not looking for more stakeholders' responses but rather seeking quality or favorable responses from critical stakeholders that matter most in helping to move the project forward.

From a technical perspective, however, the software upgrade project can be truly successful with the effective use of two key categories of metrics. These metrics include business-based or financial-based metrics, respectively. Once the business need or project objective is clearly developed, the category of metrics chosen can be structured to either support quantitative or qualitative data-gathering processes. The metrics we have listed above can be used to supply information in both categories. Quantitative metrics support critical information collection in numerical terms which can be used to evaluate business choices from a cost-benefit perspective. Qualitative metrics, on the other hand, are metrics that seek to promote value-based outcomes in project management processes. For example, improved efficiency, effective stakeholders' management, and customer satisfaction are all examples of qualitative metrics. For the software upgrade project, some of the quantitative metrics which I have highlighted to be used include (1) the number of milestones implemented according to schedule, (2) total cost per requirements and estimated cost at completion. In summary, both qualitative and quantitative metrics are vital, and they should be applied to project management in a way that adds value.

Project management office (PMO) capability development

Another important capability needed for project management success is the PMO. The PMO is a group of project-related offices within an organization that is responsible for setting, implementing, managing, and evaluating project, programs, and portfolios activities across the enterprise (Kerzner 2017). To support the effective development of a project management office (PMO) within an organization, one must first understand what level of PMO is needed and why. But most importantly, a clear assessment must first be carried out to understand the current competency level available to support the desired PMO's continuous growth and development. Once the PMO needs and organizational competency level have been identified, there are formal processes that need to be implemented to begin the PMO development process. A few of the critical steps that should be taken include

- assessing the organization's overall business systems,
- evaluating any project management capability, and
- determining the prospective PMO business purpose.

From this point forward, the most important task required is to begin with an assessment of the subject organization business profile including its vision, mission, business objectives, and organizational culture. Next, the evaluation phase should focus on examining the current project management (PM) capability to determine a clear baseline upon which the new PMO establishment functions should be developed and continuously improve. Moving forward, the PMO's main business purposes are to be determined and clearly communicated to all stakeholders to provide a legitimate justification for pursuing the investment.

Basically, there will be no quick-fix solutions even if the process is thoroughly pursued and carefully implemented. This is so because the challenges relating to inconsistencies in project administration differ greatly across organizations. Nevertheless, a combined application of the new PMO's competency along with the individual capabilities could be effectively applied through a standard approach that offers multiple avenues to evaluate standard practices and to also develop institutional mechanisms for project system improvement. An early understanding of slow but steady progress provides project leaders with the leverage to work and consistently support the PMO development process even if the immediate results are not meeting the needed objectives. We will now focus on some in-depth analysis of the key standards that should be fully understood and effectively applied when developing a prospective PMO. The analysis provided here would place particular emphasis on the development of a standard PMO. I will further explain what is a standard PMO. For now, I want you to focus on knowing that the foundational approach to be applied in the development of a standard PMO includes an effective understanding of the following key concepts:

- Organizational structure and PMO type
- Governance approach and management
- Project culture and risk management framework,
- PMO and business integration system
- PMO capacity development approach via the use of the project manager competency development (PMCD) framework

Framework for implementing standard PMO

There are different types of PMOs. However, three of the most critical used PMOs include supportive, controlling, and directive PMO, respectively. A key point to remember is that the need for a specific PMO is often derived in relationship to the PMO business's purpose or strategic project needs. Accordingly, since the focus of this effort is to provide a business case for the establishment of a standard PMO, the select PMO type best suited for such need is the directive PMO. A directive PMO implements standard control and applies consistent standard improvement across project management activities. When effectively utilized, it provides critical developmental improvement for the advancement of the project management office.

- *Directive PMO type.* As discussed, the official PMO type that best suits the establishment of a standard project management office is a directive PMO. The rationale for this choice is directly related to the organization's project management needs which are also used to support the PMO's establishment functions. In reference to the Project

Management Institute, a directive PMO works to assure maximum control of project activities and serves as the legitimate custodian of all project resources within the organization. Its fundamental functions, among other things, include but are not limited to setting standards, providing support, and leading monitoring and control vis-à-vis project oversight management efforts. Ideally, the PMO sets governance standards, evaluates organization-wide performance, and, most importantly, requires and enforces compliance for all project management activities.

- *PMO structure.* A PMO structure is made of the division of labor framework designed to support effective implementation of responsibilities. The desired structure for the standard PMO can include, at a minimum, a PMO executive sponsor, project director, project managers, and project staff members. The structure is usually influenced by the demand of portfolio size, projects' scope, or program management complexities. Whichever way the project team chooses to design the structure must simply be informed by the projected activities as well as the business strategic needs. Meanwhile, the establishment structure should first be approved by the PMO's charter. In this case, the structure includes the executive sponsor, project director, and project managers. This structure also helps to facilitate the effective application of business and project management functions so as to fully support the designated stage 3 and/or standard PMO which is being discussed as a case study in this text.

Once the PMO type and structure are fully agreed upon, the next phase is to focus on designing governance standards and policies for project management controls. This stage is critical because it is the part of the project management office framework that almost all other functionalities are based upon. The standards developed at this level can very often have deep impact on the PMO operations and overall governance. One important point to be cognizant of at this level is the project governance standards could also affect the leadership needs and project management approach. For example, if a set of rigid standards are adopted, this can automatically impact the project management culture. Organizations developing a PMO governance strategy or policies should always focus on a variety of innovative techniques to support collaborative efforts that clearly support process-based improvement. The intention of a PMO is to advance project management within an organization. Therefore, the development of governance standards should be tailored toward promoting nothing short of creative capabilities within the broader enterprise system.

Why develop project governance standards? Essential to implementing project tasks are project governance standards and project tools. From a management standpoint, a PMO with the ideal functional capability utilizes efficient governance strategies and customized project tools to deliver objectives as needed or even exceed project expectations. By definition, project governance involves the implementation of critical standards including observing compliance requirements from project managers, assessing implementation procedures for aggregate project management activities, ensuring both customers and project needs meet industry standards. Others include evaluating reports and managing communications activities across project operations. Project tools are a set of customized applications and/or software that enables organizations or project managers to perform the duties necessary to accomplish project objectives.

The PMO governance strategy is therefore constructed to facilitate the development of authorized standards to help accomplish the business and project management purposes, respectively. It is important that the governance strategies developed are fully aligned with the following critical management objectives: (1) aggregate project resource management, (2) business/project contracts management, (3) alignment of project objectives with the overall corporate strategy, and (4) development of critical standards to enhance customer value management (CVM). In the same way, a set of different tools should be arranged to help implement and monitor the necessary tasks needed to ensure the PMO governance. There are many tools that can be utilized in this area. A few of the important ones include the project charter, statement of work (SOW), work breakdown structure (WBS), stakeholders' relation management (SRM) plan, as well as a customized application and/or software to support the integration of centralized data management within the organization's enterprise management system (EMS) for project planning and implementation.

As stated earlier, the project governance standards and tools help to build and shape the project management culture. However, building an effective project management culture often requires continuous leadership development. Defining competency level is only the start of such process. Essentially, organizations seeking to improve their project management culture must be willing to invest creatively in human capital development to grow and deliver effective leadership teams needed to define and support strategic long and short-term business growth and development solutions. Research in this regard have shown that when it comes to achieving project management success, improved work ethics and change in business culture have proven to be critical components for organizations. Today, with the growing need for communication and collaboration, flexible company work culture is providing a new level of benefit that is often overlooked. Ideas like cross teams engagement create opportunities for development and growth be it in groups or one-on-one. To support project management improvement, these innovative ideas must be consistently promoted. So how does the PMO culture impact the business or project management development?

A PMO culture encompasses its primary approach to programs, policies, and project management systems development. It also includes, among other things, implementation, monitoring and control, and evaluation strategies, especially for performance standards. By extension, the collective project management culture involves the way the business functions and/or conducts its activities across both internal and external project settings. In the example we are discussing, the relevant PMO's culture would consist of all its project management programs, policies, and procedures governing project implementation within the domains mentioned by the project charter. The applicable cultural approach can also be summarized via the illustration of the PMO's project management methodology. Earlier in this chapter, we provided you with an introduction to the project management methodology. In this section, the project management methodology will be explained in relation to the PMO's culture while taking into key consideration some project management risk factors.

A project management methodology (PMM) is the official governing standard or strategy which defines how progressive project activities can be initiated and subsequently finalized within a particular organization. Put simply, the PMM for the PMO is the framework that also demonstrates its primary cultural approach to project management. A PMM includes the following five steps: (1) initiation, (2) planning, (3) implementation, (4) monitoring and control, and (5) evaluation (Daigle 2016). Collectively, this framework reveals the authorized approach through which project activities, including project work plan, project requirements, stakeholders' relation management (SRM) plan, customer value management (CVM), project scope validation, and many other applicable project steps, can be developed, implemented, monitored, and controlled. As a critical addition to this framework, two essential practices must be followed across the project management environment. These include (1) development of formal standards and (2) use of project metrics to support centralized data management as well as effective project performance measurement. While the methodical approach defined above focuses exclusively on project management, the standards and metrics should be developed according to project needs, especially relating to each project's critical success factors (CSFs). Basically, the CSFs which directly impact the project deliverables are to be monitored and measured using pragmatic project management standards.

The next step in the PMO development process is the idea of incorporating risk planning and management along each stage of the process. Project management is susceptible to vulnerabilities, be it internal or external. In this regard, it is highly essential that we develop earlier on in the project a proactive risk planning and management strategy. Risk management is a systematic approach developed and utilized by organizations or individuals for the sole purpose of identifying, analyzing, assessing, and implementing risk treatment methods. Business investments are characterized by risk and reward. However, executives and project managers must construct viable risk management frameworks to mitigate the impact of threats acting on vulnerabilities, especially those with the potential to negatively impact desired business and/or project results. To this end, a typical PMO risk management approach should be carefully integrated with each stage of the project management process.

Table 1

Risk management process	Description of risk management process
Risk identification	At the planning stage of each project, clear requirements are put in place to assess all threats that could likely impact project standards, including *time, costs, resources, technology,* and *performance quality*. The risk identification process incorporates the use of staff meetings and customized questionnaires to gather data on specific questions relating to potential risk that could impact project objectives. A comprehensive data should be compiled and documented in the PMO's risk register.
Risk analysis	At this stage, all the risk documented in the risk register should be carefully examined in line with the project key results areas (KRAs) or critical success factors (CSFs). The outcome of this process is to ensure that risk(s) identified for potential treatment do in fact possess clear threat capability to negatively impact project and business objectives. Once it is determined that some or maybe all the risks possess probable threats against desired project objectives, stage three of the risk management process is then initiated.

Risk assessment	Because the risk management process also encompasses the PMO cost management process, not all risks would be treated equally. Therefore, at this stage, risk likelihood or risk impact is carefully assessed against projects requirements. A rating system should be developed to rank each risk in order of severity. It must be noted that this is usually a *subjective process* that is often based on the subject organization strategic business values. Nevertheless, once the risks have been ranked, the PMO can then be moved to the next step of developing the risk treatment method(s).
Risk treatment	The risk treatment method includes but is not limited to risk acceptance, risk mitigation, risk transfer. Depending on the risk-ranking results, the PMO can apply either of the treatment methods. For example, if it is determined that the risk ranking and/or risk scoring falls below 20 percent, the organization could decide to accept the risk and document the rationale for such action in the risk repository. Similarly, if the risk score is 65 percent or more, the PMO could apply either the risk mitigation or risk transfer approach, depending on the evaluated cost for the risk response.

Moving forward, the next step is to focus on the PMO professional development. But before talking about the PMO continuous professional development, which is a very important part of improving the project management culture within organizations, another key area I want to emphasize, especially with regard to the PMO deployment, is the business integration strategy. Many organizations have proven to immediately experience cultural clash, especially with regard to power and leadership dynamics during or right after implementing a new PMO. This often happens because the integration of the PMO with the relevant organization's leadership culture is not always carefully executed. Even with the implementation of an integration road map, some cultural clash would still exist. However, by effectively implementing some simple but highly essential steps at the start of the PMO deployment process, the intensity of any potential cultural clash can often be methodically mitigated. Here are some of the steps that can be taken to help support the PMO integration with the rest of the organization.

To help demonstrate its full alignment with the business system, the PMO's charter and/or establishment authority should ensure the implementation of three key standards. First, a project executive sponsor should be designated to coordinate, monitor, and evaluate activities related to the PMO on behalf of the organization. Essentially, this individual must serve as the link between the PMO leadership team and that of the subject organization top management. Secondly, at the structural level, the PMO should be established as a business unit with the sole authority to support, control, and perform oversight activities for all projects within the relevant organization. This would require the use of cross-functional teams from various departments across the organization, and this process, if effectively handled, could also result in the development of increased collaboration and coordination among team members. Finally, the PMO should be led by a project leader with key knowledge of the organization's project management systems. This is vital for knowledge transfer, especially when dealing with legacy systems integration and/or transformation.

One final area of professional competency I want to address is the issue of challenges to the PMO. For the global PMO in particular, there are several underlying challenges that must be met with capable leadership if the projects and business integration objectives are to be clearly met. This is so because navigating the international business and/or project management field is often complex given the different layers of ethical and legal regulations which are often tailored to regional, geographic, national, or local business and societal needs. To clearly interpret some of the major challenges facing the global PMO, especially regarding project management integration, a conceptual model is described below. The model begins with the listing of critical PMO goals and transitions to the ethical and legal effects of international practices on the PMO goals. It elaborates on the technical and business challenges that must be met to help curb the complexities and finally ends with describing some of the potential business and project management challenges that could affect the PMO objectives.

Table 2

PMO goals	Effects of the PMO global challenges	Potential business and project impacts
Establish and achieve global project management standards	Level of human resource capability available across multiple project regions	Limited or poor intellectual capability could severely impact project operations and increase time, costs, and resources constraints
Apply institutionalized financial management practices across contracts, procurements, and other project management activities	Level of stability in business code and currency regulations including suitable and/or unfavorable conditions impacting the PMO's operations	Continuous fluctuations in the international currency exchange rates as well as unstable business code regulations could negatively impact project dates, resource allocation, etc.

Maintain formal control over intellectual property and ensure full security of all assets	Prevailing inconsistencies in business operations or substandard security regulations affecting personnel and assets	Limited technological infrastructure and weak institutional policies, especially regarding data security, would affect project's management credibility
Achieve full transparency in international trade operations including direct and subcontract management activities	Availability of and the level of access to international trade laws specifically governing contracts and business development operations	Unspecified access to international trade laws and/or redundant local policies regulations could impact contractual terms, affect project scheduling, and even influence personnel assembly for effective project management

The final stage in the PMO deployment is focused on continuous professional development. As simple as this may sound, it is perhaps one of the most critical processes when it comes to adopting a PMO. Most project management teams would very often prefer to leave this aspect of the PMO to future teams. So while human capital development is an important process, the truth is it is also a difficult thing to implement. Many development experts have argued that smart organizations often invest creatively in human resource development to meet business challenges in the present age while also preparing for the future. While these upfront investments may be challenging for organizations, especially as it relates to resources constraints, it is important to always remember that training and development are evolving, and the future cost of training and development will not be static. Therefore, an investment today could obviously be an important business solution for tomorrow. Here are some of a few critical steps that the PMO deployment team can take to ensure that continuous professional development is a lived culture within the organization.

It is important to remember that continuous professional development plays a critical role in advancing organizations and personnel competency levels. This is particularly vital, especially when dealing with the complexities of strategic project management challenges with particular emphasis on challenges relating to intellectual capacity development at the local or global PMO levels. Multiple steps can be adopted and implemented to deliver solutions with the goal of addressing professional capacity development. In the section below, we will primarily focus on professional development concepts such as (1) utilization of the project management capacity development (PMCD) framework, (2) use of professional certifications to enhance personnel capabilities, (3) implementation of mentorship projects, and (4) identification of critical leadership standards to support project portfolio management. Each of the professional development concepts mentioned above is further discussed in detail in the succeeding paragraphs.

- *The PMCD framework.* This framework plays a critical role in supporting individual and collective capabilities development for organizations. To better understand how the PMCD framework helps support professional development, it is essential to first break down the model and carefully synthesize its primary objectives. Many project management books provide detailed analysis in this regard. However, this book will simply summarize the key points. The PMCD framework, among other things, is designed to accomplish two key tasks: (1) evaluate and develop project management competency in individuals and ensure that the identified skills are transferable across industry, and (2) ensure that industry and organizations can utilize the framework to develop more industry and organization-specific competency models to advance the project management profession. Thus, the structure of the

PMCD framework works to identify units of competence, element of competence, performance criteria, and types of performance evidence. The graphical model can be applied by implementing three key steps including (1) conducting competency assessment, (2) planning PMCD implementation, and (3) conducting competency development. When the PMCD framework is fully aligned with the need for professional development, the three steps illustrated above can be used to carry on and/or support professional development within the organization and the PMO as a whole by (1) assessing individuals competency levels to determine baselines capabilities, (2) plan PMCD execution to identify elements of competency, performance standards, and types of evidence related to individual skills/abilities within the organization, and (3) conduct capability development, and document best practices to help manage continuous improvement of professional skill sets for the organization.

- *Professional certifications and requirements.* Another important step that can be taken to improve PMO competency is continuous professional training and development. Primary to fulfilling the PMO business purpose is the identification of needed professional skill sets. The PMO must identify, evaluate, and designate critical professional development competency levels to support personnel capacity development in the organization. Today, due to the level of increasing diversity in the project management field, project management professionals must lead careful market research to locate professional development opportunities that best fit their organizations' project needs. This book, however, designates three important project management certifications and their respective requirements that are particularly vital to the project management industry now and for the foreseeable future. The three certifications include (1) Agile Certified Practitioner (PMI-ACP), (2) Project Management Professional (PMP), and (3) Certified Associate in Project Management (CAPM).

Due to the critical demand for these high-level professional certifications, organization leaders must therefore consider the professional development path as a strategic business investment to be made with a careful cost-benefit analysis. The following requirements are designated for each of the certifications mentioned. PMI-ACP is one of the fastest-growing PMI certifications, and the following prerequisites are required for the exam: (a) 2,000 hours of general project management experience, (b) 21 contact hours of agile practice training, (c) 1,500 hours working on agile-type projects, and (d) masters in IT, project management. PMP is a leading global professional certificate with high recognition in the project management field and has the following requirements for its exam: (a) 36 months of project management experience, (b) 4,500 hours of proven work in project leadership with areas of expertise covering strategic planning, risk management, earned value analysis, stakeholders relation management (SRM), and (c) masters in IT, project management. CAPM is widely recognized as an entry-level certificate for professionals looking to get their PMP, PMI-ACP, and more. Its requirements include at least 2 years of fundamental understanding of project management including (a) project work plan development, (b) project scope management, (c) project cost management, and (d) communications management and/or a bachelor's in IT project management.

- *Project management mentorship.* For competitive organizations, smart investments are often tailored toward supporting both short and long-term strategic changes. To this end, the role of mentorship can be utilized from the perspective of providing the relevant organization with continuous business advantage through collective personnel competency development. Professional mentorship is the art of identifying personnel capacity needs and leading continuous effort to impart knowledge and skills with the goal of advancing performance standards of individuals and the organization as a whole. Organizations can effectively utilize this approach by making use of highly-trained internal professionals that can be used to help develop the general workforce. To better support the implementation of the professional mentorship process, a systematic approach within an organization's functional project management environment can be adopted by the PMO for personnel mentorship. Regardless of what approach is used to build the model, it should seek to incorporate four key standards. These standards include (1) establishing a management mentoring program, (2) identifying project management mentors key capabilities, (3) implementing mentorship initiatives, and (4) evaluating mentoring programs. A summary description of the model begins with the identification and establishment of organization-specific mentorship approach. At this level, strategic decisions are made as it relates to defining mentors' needs, including linking those needs with key business development priorities as well as sourcing sustainability strategies to help support the long-term maintenance of the program. Midrange level of the model involves identifying mentors' qualification requirements including professional and leadership experiences needed to support the mentorship program. At this level, concrete steps should be taken to ensure that the competency level of mentors needed to support the initiative is selected and a framework for engagement with prospective mentees are well-developed. Put simply, the key objectives of the program are defined, and a solid plan is rolled out for mentorship engagement across the organization. The final level involves the implementation of the program through engaging and empowering selected apprentices. At this stage, all necessary steps should be taken to ensure that the organization's investment in the program is efficiently managed. This goal can be observed through the implementation of a formal project evaluation exercise that seeks to assess the strengths and weaknesses of the program and also ensure that steps are taken to eliminate all inefficiencies while incorporating cost-effective measures to further improve the program objectives. Finally, it must be noted that while the mentorship program may provide a basic approach to support mentoring activities for organizations, the PMO within the pertinent organization is not obligated to implement these steps provided. Essentially, other cost-effective measures suitable to the organizational needs can also be pursued as necessary.

- *Identify project portfolio leadership capability.* Building effective leadership across project portfolio management is fundamental to improving project management capabilities. The ultimate value of efficient project portfolio leadership across modern organizations is greatly aligned with improvement of the organization's project management maturity level. This analysis has been justified by several project experts who have argued that effective portfolio management is the art of achieving excellence in project administration. A streamlined portfolio management process incorporates steps that are simply the fundamental activities of the PMO's portfolio actions. However, this level of competency can only be supported and successfully sustained through increased project management capability improvement. The emphasis in this process is placed on associating success with the continuous advancement of the PMO. This is achieved via improvement in project management maturity. Today, many project management experts have argued that effective portfolio management is only possible through the identification of performance weaknesses at the individual and aggregate project levels. This is essential for building a bottom-up leadership project management strategy at the PMO. However, the data collected is not simply limited to meeting leadership needs. It can also be evaluated to support the development and implementation of techniques required for standard portfolio management operations at the strategic level. This works best when major weaknesses and strengths in project leadership have been identified at the baseline level. By achieving such level of competency, critical management qualities such as advanced project leadership and technical business management can therefore be developed to support this effort of cultural transformation from project to portfolio levels, respectively. An important way to support improved portfolio leadership management is to take steps that specify the critical need for developing the PMO competency improvement at all times. This can be done by analyzing the different stages of the PMO development and taking delicate steps to plan for and improve identified inefficiencies where necessary. To successfully implement this process, a clear understanding of the PMO stages and functions is required. In the below paragraphs, this text provides a brief description of the PMO development stages.

- Project office—provides progress reports on all projects to the higher level PMO for the purpose of evaluating project data relative to the business strategic needs.
- Basic PMO—introduces fundamental concepts for portfolio management through developing processes and criteria for project selection and alignment with the strategic objectives of the organization.
- Standard PMO—establishes and manages collaborative processes to include a centralized operations management for project portfolio data. This level of the PMO usually eliminates the activities of isolated project data management while leading efforts to support aggregate data collection and project portfolio analysis and improvement.
- Advanced PMO—creates comprehensive portfolio management capability, establishes a portfolio management review board, and employs systematic effort to facilitate real-time data collection and effective portfolio decision-making.
- Center for Excellence—incorporates and reviews the activities of lower-level PMOs and develops policies and

guidelines for centralized portfolio management. Creates organization-wide standards to support and facilitate portfolio management and employs formalized management review strategies to monitor and evaluate business and project inefficiencies related to portfolio management. By understanding the stages and PMO functions, competency baseline can be established, and clear steps can be taken to improve project portfolio leadership.

To summarize, we must remember that the core focus of the PMO is to ensure that any current level of disorganization within an organization's project management environment is effectively addressed. By establishing a PMO, organizations, project managers, and their respective teams would eventually be able to adopt new and advanced project management measures to facilitate strategic and operational improvement across project development activities within the organization. As discussed also, actionable steps should always be incorporated under the PMO professional development section that can be used for the sole purpose of supporting progressive personnel training and development including improving technical, business, and project-specific capabilities.

CHAPTER 4

Promoting Ethics in Leadership and Technology Management

Across chapters 1 through 3, we focused on concepts relating to technology and leadership integration. Our goal in those chapters was to provide awareness of and deliver key informational resources to help breach the inherent technical isolation (ITI) culture. In practice, ITI is a management culture adopted either actively or passively by most organizations where nontechnical personnel or people in general are consciously or subconsciously isolated from technology processes due to some perceived limitations as it relates to their understanding of the subject matter. This is a culture I believe we need to drastically change. Today, technology is not simply an abstract idea. Technology has profoundly impacted humankind for centuries and will continue to do so in even greater ways for the foreseeable future. It has changed our way of life, impacting almost every area from homes to work, from school to businesses, and has even impacted our spiritual practices. What this means is that technology can no longer be treated as a concept reserved for some special people. In short, it must rather be understood, shared, and pursued with the key objective of delivering a general good for all. In chapters 4 and 5, I will focus on ethics in technology and the adoption of technology solutions. We will build on previous efforts to give you, the reader, a clear approach regarding ways to not only understand technology but also develop the skill sets to effectively use, adopt, and subsequently recommend progressive changes to tech applications. This chapter will focus on ethics and leadership in technology management.

Ethics in business management

Ethics is one of the most debated subject matter in technology today. The methods we use to conduct our businesses to improve innovation and deliver effective customer service have a clear impact on the larger environment within which we operate. To this end, ethics plays a vital role, especially when it comes to shaping business cultures to meet new environmental, social, political, and technological demands. Some will even argue that ethics has made business management more difficult. In this text, we will look at how ethics has shifted or continue to shift business cultures in both technology usage and leadership development.

In modern business management, workplace ethics is a popular concept. Today, many organizations take great pride in the development of strong ethical values that are used in guiding the business course of actions. Nevertheless, workplace ethics emerged out of a history of human struggle. An interesting historical perspective reveals that ethics was never always present in organizations. Some research has it that in the US, for example, ethics made its way to the top of corporate institutions because of a fundamental shift in cultural values in the late 1940s. As the change in political culture began to shift more toward public and private sectors' accountability, the concept of ethics stood out. In the early 1950s, the growing popularity of labor movements followed by the continuous debates about social welfare and economic inequality laid the foundation for modern workplace ethics. Some leaders have also argued that by the end of that decade, President Truman's fair deal, which dealt with issues such as fair income, civil rights, and business social responsibility, had a major influence on pushing ethical behavior into the agenda of corporate America. However, it was the willingness of some organizations and business leaders to take a chance on implementing ethics that really made it a possible cultural revolution.

In the era mentioned, as organizations pursued new ways to inject the culture of integrity into their business operations, the concept of workplace ethics became a key subject of corporate interest. Today, many organizations emphasize ethical standards such as integrity, professionalism, accountability, teamwork, collaboration, and commitment as key corporate values. Some managers train and hold individual employees accountable for living up to these standards. Accordingly, it is a well-known concept that somehow, the modern demand for business integrity has forced many organizations to now consider workplace ethics as a fundamental part of the organization development process. But what is workplace ethics, and how does it affect the average employee? Let us look at a few important concepts in ethics before discussing workplace ethics in detail.

First, it is important to understand that morals, ethics, and laws are defined differently and could be influenced differently across societies and cultures. *Moral* is defined as an individual personal belief about right or wrong. *Ethics*, on the other hand, is an agreed-upon standard or set of beliefs about right and wrong behavior within a society. Ethics differs from morals in the sense that it is usually a set or a code of conduct developed by an organization or group with the intent to guide individual and/or professional behaviors. *Laws*, as we know, represent a system of rules that tell us what we can and cannot do. We will now move forward with the discussion on ethics in the workplace. In the latter years of the ethical revolution, as management evolved, new studies about workplace ethics were being developed, examined, and marketed across the business ecosystem. Today, there are many innovative strategies on how to improve workplace ethics. Below we discuss the six key approaches that business leaders can adopt to systematically promote and sustain the practice of ethical behavior within their organizations, especially as they seek to improve and manage technology processes.

Appoint a corporate ethics officer. The first of our six steps is to appoint an ethics officer. According to most experts, having a corporate ethics officer provides the organization with the vision and leadership needed to develop and support ethics program management. When organizations appoint ethics officers, it represents a statement of intent to develop and promote the culture of ethics. With an established authority, the corporate ethics officer can help cultivate the organization's talents and work to design ethical standards to be utilized by employees across the entire organization. In fact, with the appointment of this officer, some of the key activities that can be carried out include ensuring organizational compliance, creating and upholding ethical standards within the organization, and finally, the officer also serves as the first line of contact and knowledge source for the implementation of the organization's ethics affairs.

Develop ethical standards. With the appointment of a corporate ethics officer, a framework can be put into place to develop ethical standards for the relevant organization. In most organizations, the board members are in a unique position to set clear examples for the promotion of ethics through practicing and choosing to act with integrity at all times. These leaders can build a culture that supports the highest ethical standards by leading by example. Interestingly, not all board members would usually endeavor to set such standards. To this end, it is interesting to note that when board members fail to conduct themselves appropriately, the impact of their actions often sets a negative example for the rest of the organization. This could even lead to employees' discontentment and can sometimes possibly promote a radical revolt in organizations. However, when board members implement strict ethical practices and live up to what they preach, executive members and employees get the message and follow accordingly.

Establishing corporate code of conduct. Another important step that can be used to promote ethical behavior across organizations is the establishment of standards to govern and control business development actions within the organization. The code of ethics is usually a formal statement that specifies the purpose of the organization, its values, and the principles that should guide its employees' actions when implementing operational tasks. Usually referred to as a statement of value, the statement specifies the business approach to integrity when dealing with internal and external affairs. By setting a clear code of ethics and effectively following such standards, organizations can help to improve ethical behavior across their business environment. The code of ethics directs the actions of board members, executive officers, and employees of the organization. Its major objective is to ensure that everyone acts according to clear guidelines with an understanding that there are consequences for any breach thereof.

Conduct social audits. Moving forward, managing an organizational ethics process should not be static. Business leaders must develop creative ways to promote ethical actions at all times. Another key managerial activity that promotes ethical behavior in the workplace is social audit. Social audit is a process by which organizations evaluate how well they are meeting their ethical and social responsibilities. The intent of this approach is to carry on fact-findings and establish a baseline for the organization's future course of action. This step, when effectively applied, helps to promote a sense of value and accountability around the organization's ethics goals. Today, some organizations even share their social audit report with relevant external stakeholders to help promote interactivity and transparency. The social audit report reveals the organization's performance in relation to its ethical and corporate social responsibility. It also helps the management team to be aware of where they are as they seek to set new goals for improving ethical relations across the business ecosystem.

Implement ethics training for employees. Continued education is one of the historically proven methods that lead to improvement in workplace ethics. Therefore, implementing customized ethical training programs should be a vital part of the organization's drive in promoting ethics. The organization's code of ethics must be consistently practiced and frequently communicated to employees at all times. Activities such as ethics workshops involving historical and contemporary case studies on ethical decision-making processes must be presented to and studied by individual employees through frequent ethics training programs. This will improve employees' conduct in relation to the organization's code of ethics. I cannot overemphasize that constant training leads to improve ethical behavior within organizations and can also boost a company's ethical ratings.

Include ethical reviews during employees' appraisals. Lastly, organizations must also include ethical evaluation as part of their employees' performance evaluation process. Some organizations that have implemented this step have instantly reported improved ethical relations on the individual and aggregate levels. Put simply, whenever an organization decides to evaluate employees' ethical behavior, it improves accountability around ethics. Some of the critical areas that can be looked at include what drives motivation in a multicultural environment. How are people accepting personal accountability for meeting business needs and taking responsibility for developing others? A key thing to remember is that this process often explicitly sends a message throughout the organization indicating that management is fully in support of acts demonstrating the highest ethical standards. Moreover, by integrating ethical standards with traditional methods of evaluation, organizations tend to experience improved ethical actions from their employees over a period of time.

Five critical steps to creating an ethical model

We have talked about some of the direct actions that organizations can implement to help promote ethics in the workplace. Many of the actions can be adopted from a variety of knowledge management resources across organizations. However, given the increasing complexity of ethical challenges, it is important for organizations to begin thinking about creative ways to adopt or develop ethical standards not only from outside sources but from internal sources as well. Creating best practices around ethics can also be developed internally via an ethics model. This process, while often difficult to implement, represents one of the ways that can truly promote the ownership of values around ethics while effectively helping to drive an organization's ethical processes. One of the key reasons why it is important to build an inclusive ethical model from within the organization has to do with the controversies that often arise from ethical issues relating to workplace monitoring.

In most companies, the goal of workplace monitoring is to collect real-time data on how employees use work time, company equipment, and other facilities placed at their disposal. Essentially, companies invest a lot of time and money in hiring and maintaining employees, and as such, getting the best out of their employees is usually a top corporate priority. Whether utilizing internal or external hiring teams, activities relating to setting up career events, posting on job boards, conducting interviews, and onboarding of employees make hiring a costly and sometimes a very difficult process. For example, some companies have suggested that when you total all the necessary costs to hire and retain an employee receiving a salary of about $50,000 per year, it is likely that the employers pay anywhere between $60,000 to $70,000. Take nothing away from dedicated employees; however, when employers look at rising payroll costs, it is natural that they will stop at anything but getting the very best out of their employees even if it means dedicated monitoring of work activities. To this end, building an inclusive ethical model is very important. However, given the issue of privacy concern, a big question for employers to answer today is, How does workplace monitoring come into conflict with employees' privacy concerns?

From a technical perspective, there are two major issues with workplace monitoring. First is the issue of how it impacts *work culture*, and next is the *privacy concerns* regarding data collected about employees' professional behaviors. In some cases, when workplace monitoring becomes extreme, it engenders a culture of fear and limits flexibility, especially with regard to natural human interactions within the work environment. Some ethics leaders have even argued that while effective workplace monitoring may be good, the psychological implications it sometimes has on the employees could render the whole initiative provocative and even disastrous. Thus, approaching such a policy with an inclusive ethical spirit across the organization can be very important for achieving optimal results. When employees fear that constant management monitoring is impacting their personal productivity, it sometimes reduces the communal coexistence within the company and can impact progress to the point of even inducing personal resignations.

It is therefore very important for companies to balance the need for effective workplace monitoring with the goal of promoting an open and friendly work environment. As mentioned, there are also concerns regarding what companies do with the data they collect on employees' behaviors. While management may use the data to fire, reward, demote, or implement departmental reshuffling, from employees' perspective, the concerns are however associated with the long-term security of the data. For example, several employees fear that given the sensitive nature of these records they could easily be sold or hacked by malicious internal or external actors. And this concern of insecurity relating to privacy is something that can definitely impact the professional lives of millions of people in a negative or positive way. While it is true that employees should expect less privacy at work, employers must now be cognizant of the fact that the far-reaching effects of possessing such sensitive personal records could come with weighty consequences, particularly in the form of potential lawsuits from disgruntle or perhaps heavily conscious employees, especially if there exists any unmindful handling of such records. This is just one reason why building an inclusive ethical model is important in today's work environment. Below I have documented a few steps that organizations can use to build an effective and inclusive workplace ethics model.

Assess ethics culture. Developing a technical model is something that most organizations would prefer to hire external experts to do. However, in the case of ethics, it is hard to determine what organization or who are the real experts. In fact, most people who strongly believe in promoting ethical values think that the only expert solution in ethics is the standard we define to govern our ethical actions. Nevertheless, this belief does not in any way limit an organization from seeking external consultancy when it comes to developing an ethics model. This book, however, is encouraging organizations to begin to practice ethics as a fundamental part of the organization's development culture. To begin the process of developing an ethics model, a questionnaire should be developed to assess the practice of ethics within the organization. Remember, we have talked about what ethics is and how it is different from morals and laws. The idea of assessing ethical practices can begin by looking at something as simple as what the views of your employees are on right or wrong within the organization. Eventually, other key areas in the questionnaire can be tailored to gather information on the following issues:

- What personal values do your employees subscribe to the most, and why?
- How do employees think they complement the company's values, and what challenges do they have with expressing themselves in said direction?
- What are the most important business needs to employees, and why?
- Apart from the ethics officer, who would be the best person to talk to within the organization when moral conflict arises, and why?
- How are your company's ethical standards complimenting the workforce's general ethical beliefs?
- How do your company's ethical values differ from the employees' values?
- What recommendation(s) do you have to improve and help practice ethics more effectively within the organization?

These are simply general questions that can be incorporated into the ethics questionnaire. One important thing to take into consideration is the fact that the actual survey should be developed in line with what the organization is trying to achieve strategically, especially with regard to any moral and social implications need(s) that should be addressed to help the organization achieve its goals. A key thing to remember is that most ethics advocates have agreed that whatever actions we take to benefit not only our organizations but the community in general is a good approach. However, the model should also seek to understand how such actions can be effectively implemented amidst resource constraints. This includes looking at new laws on ethical behavior and how those laws directly or indirectly impact the organization's business processes.

Define ethics baseline. Regardless of how the questionnaire is developed, the key objective is to come up with clear data after the assessment process. Basically, at the end of the process, the implementing team should have a reasonable data relating to where they are in relationship to the practice of ethics within the organization. From this point forward, evaluating the data and deciding what to incorporate as baseline standards against which ethics codes or actions would be developed is key. In the next stage, most organizations would try to base the development of codes on what the state or other governing authorities have already defined. What we recommend, however, is to look beyond such thought and think within the larger context of becoming an ethical leader.

The goal is to not only limit the ethics plan to what your state or governing authorities have defined but to focus on developing innovative ways that can even push these authorities to place greater value on ethics. This is sometimes considered as an act of redefining or leading a business ethical revolution. For example, a manufacturing organization must not only seek to stay within the limit of greenhouse gas emission as prescribed by the law; they must seek creative ways how to collaborate with communities as it relates to finding alternative long-term solutions where communities can be served and business needs can still be pursued without creating environmental impacts that are detrimental to the current and future generations. A collaborative leadership process can be developed to help businesses and community leaders work around such an idea. I must remind you that this is not going to be easy. However, times without number, history has proven that when we work together, we are definitely capable of tailoring our development processes to meet the most critical challenges that were previously deemed unachievable.

Develop ethics standards. Documents from the ethics cultural assessment process should be carefully reviewed with a clear objective to support the development of standards. A unique change at this level is that the model seeks to support active analysis and careful transformation of data into information to clearly inform the ethics guidelines development process. After the data analysis, four critical steps are taken to begin developing the standards. The steps include (1) specifying project scope, (2) defining ethics goals, (3) determining data collection sources, and (4) data analysis and standards development. It is important that activities at this stage are organized in a way that, from time to time, there is frequent communication between the project team and the rest of the organization. Developing an effective model to help continuously produce ethical standards that help promote best practice requires consistent and highly effective communication and collaboration across the organization.

As you continue reading, many might have already asked this question: Why develop internal best practices on ethics? In 2007, many loopholes relating to business ethics and financial laws were aggressively exploited to the detriment of the general public. We all saw the effects of the financial crisis, but how many of us could have argued or perhaps argued that companies will need to do more in challenging themselves to build high-quality internal best practices on ethics? The question is, Was the financial crisis a weakness of the law, or were individuals just applying exploitative moral judgments? However you answer this question, just bear in mind that most companies are truly driven by the values they aspire to live up to. To this end, if there exists a continuous degree of moral deficit, no amount of laws would change the attitude of big or small corporations. This is why recommendations such as conducting effective social audits on ethics-related issues must be carried out from time to time within the organization. Therefore, an internal model is important to set standards for individual and collective accountability.

Implement ethics standards. Developing and managing an implementation framework for execution of ethical standards is one of the most important parts of this model. At this stage, once organizations have defined and clearly linked the ethics goals to the overall strategic vision, strategic ethics milestones can then be defined. These milestones are a function of the ethics critical success factors and must be monitored and measured with effective metrics. To put a grip on how organizations can help their teams experience and effectively practice proactive implementation, a combined management approach of communication, training, and standards integration with mainline business processes is highly recommended. Do not send out an email on ethics standards and tell people that a copy of the document is available in the knowledge resources room or at this link. Instead, make this document a living initiative within the organization. Put simply, engage with team members to help consistently promote these values at the strategic and operational levels.

The email is not enough. Essentially, how ethics is practiced matters more than how it is communicated. For example, some organizations set a goal to value all their employees, but the reward system is based on who knows who. In this example, the value contradicts the practice which, in turn, demonstrates the limitation of mere communication. If your ethical standards regarding rewards favor some team members over others, you simply open a back door for discontent that could even lead to a possible rebellion. The focus of leadership is to sometimes encourage competition and reward; however, if the emphasis is placed on the competition, we usually miss the point. Valuing the human being is perhaps the most essential act that can be taken to promote effective leadership at every level within the organization and beyond. When people first feel appreciated for who they are and not what they bring, you spark a light in them to challenge themselves to live up to a higher calling. Do this, and build leaders.

Evaluate and improve actions. At the end of every growth and development journey, a moment of reflection, stock check, personal evaluation, or whatever you call it is required to help assess actions against predefined goals. Under the model, the evaluation steps are set in accordance with understanding how key results areas were arrived at, what critical success factors supported the key results areas, and finally, what results we have based on the actions we took. The evaluation idea is to simply ensure that organizations seek to appreciate what was done right while working to improve things that can be made better. It is not, however, a quick-fix process. The evaluation exercise can go for weeks or months, depending on the complexity of the task or tasks outlined in the process. Organizations can use one-on-one meetings, group engagements, anonymous surveys, or whatever method applicable to their settings. In the end, evaluation results should be used not to emphasize deficiencies but improve application effectiveness.

Today, with the use of advanced metrics, ethics leaders can be able to not only have continuous access to potential evaluation results but can also begin taking advanced steps on what actions could be needed to positively impact desired outcomes. This is the act of being proactive. This is the benefit of leading metrics—they are proactive information sources that can be used to build real-time responses to critical and noncritical business challenges. The ethics team plays a key role in this process because it is their responsibility to not only share this information but to break down actionable steps on how negative indicators can be transformed positively. Again, as you have already repeatedly read in this book, the impact of effective leadership on driving the use of technology cannot be overemphasized. Getting the best value simply comes down to effective process management. Therefore, the leadership process in managing not only our evaluation exercise but also determining improvement techniques must be clearly defined.

Ethical constraints and privacy challenges in the technology age

Ethics and privacy represent some of the most controversial topics heavily debated in today's technological age. The increasing demand for advanced technology solutions coupled with the growing complexities of business and security needs often place proponents on different sides at the edges of their seats when it comes to policy focus and implementation. Unlike many other controversial subjects where there often seems to be a reasonable balance, it almost always appears like it may take a lifetime to arrive at any reasonable resolution on this matter. Today, with cybersecurity warfare increasingly becoming a threat to nations, conscious individuals are getting particularly concerned about how private data stored by government authorities can be truly protected against internal and external cyber threats. With constant breaches, scrutinizing institutional credibility, especially when it comes to citizens' private data protection, will remain forever under the spotlight.

This book does not attempt to provide any resolution on this matter but rather expands the debate on the critical need to seek effective collaborative leadership in bringing urgent solutions to the privacy and ethical challenges we face now. In July of 2021, United States President Joe Biden made it clear to US intelligence agencies that persistent cyber threats might end up leading to a war with guns. The president was referring to the rapid cyber breaches that now appear to be completely out of control. According to the Information System Auditing Control Association (ISACA), there is a new breach at least every thirty-five seconds. DevSecOps, a term that simply means incorporating security from the start to finish in every software, is now a path within the technology landscape that cannot be overemphasized. Despite utilizing DevSecOps, organizations are encouraged to continue to pursue other important security management techniques like network scanning, penetration testing, and security information event management (SIEM).

What are some of the privacy and ethical challenges we face right now? While some countries have passed major laws in favor of privacy rights, the effective management, protection, and impartial execution of such laws remain critical areas of concerns. Whenever we think of privacy in the simplest term, we think about the right of an individual to conduct his/her life without anyone else's interference. This definition may not fit all possible scenarios. However, let us keep our focus on privacy and ethical challenges in the context of modern technology. Today, modern technological innovation is changing the way we look at privacy and, we can say with certainty, is truly impacting individual privacy in many ways.

Across the industries of business, entertainment, government, politics, and religion, the availability of advanced communication tools that can be used to promote collaboration, effective communication, and efficiency in operational services has changed the way organizations collect, store, and manage private data. Today, to facilitate online transactions, for example, both employees and customers use various online systems which sometimes require personal identification setup information such as usernames and passwords. The ethics of how such information is preserved and protected without any potential privacy threat to the users are part of the credibility challenge facing several organizations. Inefficiencies across private data management and adequate protection are raising more and more concern for amendments to internet privacy laws.

Many people believe that the right to privacy is a human right. Citizens of the United States, for example, are guaranteed privacy through the Fourth Amendment to the US Constitution. Elsewhere, after experiencing several breaches, more and more new laws are being passed in the British Parliament to highlight how UK citizens can benefit more when it comes to privacy rights over personal data. In Africa, Article 16 of the Liberian Constitution guarantees the citizens' right to privacy. And these are just a few of the many countries with organic laws on individual rights to privacy. However, advanced digital surveillance and increased workplace monitoring are perhaps the most critical areas of ethical concerns when it comes to individual privacy rights.

Across different societies, various instruments including laws and policies are used to guarantee privacy rights. Yet privacy in modern times is a complex issue. Many times, the possibility of domestic and foreign threats influences our decision-making about privacy. Notwithstanding, we cannot bury the importance of privacy. Some thought leaders in ethics have suggested that new and advanced surveillance technologies that have the potential to capture information about people's private lives should be relooked at. Interestingly, the people behind these cameras can execute their surveillance mission with or without the consent of private citizens. Even though advocates for these technologies emphasize the positive side such as threat prevention and crime control, such activity, many believe, sometimes erodes the very foundation of the individual right to privacy that we claim to protect. Going a step further, the act of persistent secret surveillance continues to leave many people feeling insecure, especially as it relates to what is being done with their private information.

Also, the exponential deployment of advanced surveillance cameras across major cities is forcing critics to question both the cost of the equipment as well as the underlining objective of the action. The objection from most people is not to counter public safety but rather to seek assurance of full privacy, especially when hackers are now breaking into schools, gas stations, traffic cams, and many more to steal targeted individuals' personal data. Critics fear that private records could be breached and used to blackmail individuals. They also argue that such activity not only intrudes upon an individual's right to privacy but also creates room for fear and slowly impedes freedom of movement. There are some individuals who would now deliberately choose not to ply a route due to extreme public surveillance. In addition to several ethical concerns about the management of private citizens' data, critics strongly think that continuous advanced surveillance will limit freedom of movement. Put simply, if I have a camera on me everywhere I go, I may live in a free state, but I am simply not free.

Another key area that often poses ethical and privacy challenge is management monitoring of employees. Employers' roles and responsibilities can sometimes conflict with workers' right to privacy, especially when employees feel that there is a reasonable room to assume privacy breaches due to extreme management monitoring. For example, in 2012, the US Food and Drug Administration (FDA) admitted to monitoring the private email accounts of nine of its scientists who had expressed concerns about the FDA process for approving medical devices. For its part, the organization sent out a conflicting message to employees by informing them that while their email may be monitored, limited personal use of their government-issued computers was acceptable. Such a message could be interpreted as setting a reasonable expectation of privacy. In the end, the affected employees filed a lawsuit claiming that the FDA officials violated their privacy and constitutional rights by monitoring their private email communications. I will not delve into the full story on this case; however, I urge you to go out and research how it ended and what questions were left unanswered about workplace privacy rights.

Another great example was articulated in an advanced workplace ethics research done by G. A. Prout. Prout argued that over our lifetime, employees provide highly valuable private information to employers. Many of the information, he said, form the core of our life activities including date of birth, social security numbers, phone numbers, mailing address, marital status, previous job history, kids and spouses, and travel history. He explained that given the sensitive nature of these pieces of information, most employees sometimes find it difficult to release them to employers but would have to due to job needs. The organizations then take the responsibility to gather, review, and file these pieces of information in order to help with processes such as employee background check, professional history, taxes, legal status, and hiring decisions. He concluded by saying that many employees do not feel comfortable releasing such private details to anyone but admitted that we have to do it since it is required by our organizations. This, he maintained, is a perpetual source of privacy conflict between employers and employees.

As I mentioned earlier, this book is not seeking to provide any definite resolutions on these ethical and privacy issues but rather to bring them to your attention. The goal is to encourage effective leadership around these issues so as we move forward, the better it gets for all parties. We all want to be protected, and if this is the goal, a common ground must be reached on applicable standards to support the ultimate privacy goal. The right to privacy is, with no doubt, an important aspect of human coexistence. Even without admitting it, many people just want to be left alone. What they eat, where they shop, the type of clothes they wear, and how they spend their private time is of huge importance to them and their families. For many people, carrying out these simple but delicate life activities without the fear of being constantly monitored is a priceless dream. But in today's technological age, it is difficult to imagine how people could go about these activities without possible monitoring. Surveillance cameras are posted at schools, churches, restaurants, hospitals, workplaces, and even homes. Surveillance advocates reference public safety as the number one reason for their action. Wherever the claim and counterclaims on privacy rights end, a great number of people, including myself, believe that the real solution lies in information transparency.

Incorporating ethical practices in technology development

We have talked about the different steps organizations can take to improve the culture of ethics. Also, we provided a model to help create ethical standards with a focus on team members' inclusivity. A brief analysis was also provided on how ethics and privacy are now major issues in modern times. Across these discussions, we emphasized the need for leadership and effective cooperation across teams to support needed ethical actions. However, with respect to technology development, there is a need for organizations to incorporate effective development standards and/or business strategies that combine or integrate clear ethics at every stage of software development. This is important because wide research has proven that actions such as personal or collective carelessness, abuse of authority, violation of professional ethics, the lack of integrity, as well as the misuse of corporate privileges are often very common across organizations, especially when it comes to software development project management. Here are some steps that can be taken to mitigate such actions.

Promote people over products. In 2017, a friend of mine and I worked as project assistants on a marketing project for a Fortune 100 company in Denver. The top management had set a truly ambitious campaign goal of increasing sales by 75 percent. To achieve this, the project manager was fired and replaced with someone who was at least considered more experienced and fully ready to lead the project. The new project manager came on with a lot of credentials. On paper, no one could argue against his appointment. At our first project meeting, he greeted everyone calmly and said, "Gentlemen, let's do this." He hit the floor, took up his marker, and began drawing the product commercialization process. He talked about creating a direct marketing plan, an advertising plan, a communication plan, a media plan, a pricing strategy, and a distribution plan.

Everything seemed to be going well until the third week into the planning stage. By week three, we experienced a sudden drop in attendance. Several key members of the project began giving excuses, highlighting other functional priorities that were stopping them from attending project meetings. Things got worse when the VP for project management, who was the sponsor of this project, suddenly resigned and left the company. When we failed to get the full project team to convene, I approached the project manager and said, "Sir, is there a possibility for you to take some time off and schedule a one-on-one with all eight project team members and try to gather sincere feedback on our current approach?" To my surprise, my request was accepted. And not only was it accepted but I was also asked by the project manager to attend all the one-on-one meetings with him. Long story short, the general consensus from the meetings showed that team members felt isolated from the rest of the project leadership team.

Two weeks later, there was no change in approach, nor was there any major attempt to get everyone back on the team. In our last meeting, I vividly recall about three or four of us sitting, and the project manager came in a said, "Gentlemen, professionalism is key to success." My friend and I did not end that meeting, and two days later, both of us left the project citing management inconsistencies, especially in the areas of communication and leadership. Six months later, we were told that the project manager was fired because he tried to introduce an employee-focused motivational program across all project management activities. He wanted a new approach to project management, one that would place greater emphasis on team relationship development. The top management thought that he had already failed to deliver, so he was quickly relieved of his role.

The simple challenge which this company faced was its inability to remove extreme emphasis on product and processes in the commercialization process. This story is simply the tip of the iceberg as there are many others whom I have spoken with who have expressed similar challenges with their organizations. Put simply, most companies today place emphasis on product over people in the commercialization process. Whenever you leave people wondering as to what value or human connections they bring to teams and processes in project management, you immediately create a formal recipe for failure. People are people, and no matter what processes they find themselves in, they will need leadership that gives them value while seeking to bring out the best in them. This is also true for technology development. When we are developing software, the focus should not only be placed on extreme commercialization in relationship to sales and end users. Let us seek to value our teams and create strong and better connections with what they create not only for sale but to serve greater human and environmental needs.

Do not suggest; stand up for safety. I am a stickler for reading business news and articles, especially when it comes to finding out what companies are doing to grow and stay competitive. However, I do not only read the news. I also like to talk to small business owners and gather perspectives on what matters most to them when it comes to growing and effectively managing their businesses. Some of my conversations with business leaders are very casual, while others are often intense or somewhat focused on methods of proven business development models. Over my ten years of business experience, I can count the number of business leaders that I have spoken with who have said to me that their employees' safety comes first. As a business developer, I do not seek to judge anyone's approach on this matter. However, I want to point out a gap that I believe is a vital missing link that so many companies are underperforming in the business world today—emotional, and physical safety.

On the contrary, one of the areas we often pay huge attention to is the development of safety-critical systems. Technology systems with any inherent defects that can lead to personal injury or death are referred to as safety-critical systems. Safety-critical systems are essential not only to businesses but to humanity as a whole. The impact of these systems on the environment is not only measured in terms of cost and profit analysis but extends across the general livelihood of human well-being, affecting communication, health, defense, family relationships, and, obviously, public safety. Many times, the intentional efforts we apply to develop these systems can really not be compared to preparing for and promoting employees' emotional and physical well-being. While some companies are making important strides in safety, a lot of organizations are still lagging behind. In 2017, the federal Occupational Safety and Health Administration (OSHA) reported that one in five deaths were due to accidents in the construction industry.

The construction industry is undisputedly a major contributor to economic growth and development around the world. In the United States, there are over 680,000 construction employers with nearly 7 million employees who create almost 1.2 trillion dollars' worth of structures annually. The growing rate of death in this industry worldwide is increasingly becoming a critical concern for not only employers but governments, customers, and employees themselves. But while we often tend to focus more on physical safety issues in the construction industry, there is a bigger concern regarding emotional safety across businesses. In 2017, *Small Business Trend*, an online news magazine, reported that awareness, stress management, and effective communication were critical safety practices that all organizations needed to adopt and effectively apply in the process of implementing safety guidelines. Interestingly, the article also stated that companies should make great efforts to begin moving toward effective safety models that focus on emotional safety just as much as we place extreme emphasis on physical safety.

When I spoke about safety-critical systems earlier, my goal was to draw similar attention to emotional safety concerns. We have been able to design some highly innovative physical safety models. However, we need to begin paying attention to our greatest assets—employees. Wide research has proven that highly competitive organizations are not just powered by the best business strategies but are also led by the most emotionally stable personnel. Essentially, there are several things that can happen when we not only communicate physical and emotional safety as policies but also integrate them into the company culture in a way that represents our strategic business values. Put simply, safety challenges must be treated as a red flag operational risk issue. At the strategic level, operational risk impacts business growth and at the implementation level. It stifles process efficiencies while limiting business innovation. When we treat employees' safety as a matter of value, we promote improve business intelligence and continuous innovation.

In emphasizing safety as a value, organizations must seek to consistently build creative approaches around addressing three key areas including awareness, stress management, and effective communication. When we develop safety guidelines to meet state, national, or global best practices, we must always remember that policy availability is not the same as policy success. It is good to have the policy documents, but it is better to keep the concept of safety alive by executing operational processes that are specifically designed with the intention to promote safety as a strategic value. Thus, the implementation and overall management process must provide employees with the opportunity to live, act, think, and breathe safely, knowing fully well that increased productivity at every level requires healthy bodies and positive minds.

Dismantle inherent technical isolation. For the past ten years, I have had the opportunity to work with several industry-leading companies in the United States. Like most employees, I have come face-to-face with issues related to ineffective IT communication culture within organizations. I have also read books that have detailed some of the challenges that exist between IT departments and the rest of the organization. In 2020, Global Knowledge concluded that workload, cybersecurity, and skill gap are the top three challenges facing IT departments today. The report emphasized that increasing workload is causing most IT departments to replace training time with working to meet backlogs. Essentially, the big challenge relating to how IT departments could work to narrow skill gaps internally is still a major issue needing critical attention. While many of these challenges are true, there is another important challenge that is not very often talked about. And even when it is talked about, many people often tend to limit the real problem to the lack of effective communication between IT departments, management teams, and the rest of the workforce.

In 2017, the project management institute reported that 14 percent of IT projects failed across many of the world's leading companies. However, that number was only for projects that failed outright from the start. But here are the interesting stats. Out of the projects that did not fail outright, 31 percent never met their goals, 43 percent exceeded their initial budgets, while 50 percent were late or completely underdelivered. Many of the pitfalls were attributed to inaccurate requirements, estimating challenges, and continuous scope creeps. I talked a bit about these challenges in chapter 3, so I will not be repeating anything here. However, what has been consistently ignored over the past years is the inherent bias of IT leaders in training and getting the rest of their teams (technical and nontechnical) on board with the overall company technology vision. There are multiple statistics out there showing failed project implementation due to improper workforce preparation.

I do not intend to bore you with stats around such issues. However, I want to rather talk about the missing link. In most companies, when a technology project is being designed, it is often rare to see a clear communication plan relating to how the project, tools, or services will be effectively communicated. In fact, using the workload and backlog clearing challenges as the major priorities, most companies no longer have the time to incorporate creative communication planning in their IT projects. Most IT managers would prefer to send out emails and links with recorded videos in hopes that these efforts will replace real communication and effective workforce training. Most business articles have even listed workforce development as the fourth or fifth priority for companies with other issues like market research, new customer acquisition, new technology adoption, profitability planning all coming in front. Yet this is not the real problem. The real problem is built on how IT departments efficiently utilize their training and communication services across the organization.

In 2014, Van Lines Inc., a Fortune 100 company working in the transportation industry, wanted to make a change in its technology culture. The general manager had noticed a huge drop in workforce productivity as well as an increased demand for IT services. Like most IT departments, the service desk was sending out at least three emails a day with links included at the bottom encouraging employees to acknowledge that they have read and understood the various issues. The problem with this process was that every week, the IT department would receive several emails from employees, especially nontechnical employees, relating to questions about policies and procedures. Instead of addressing these questions impartially, the IT manager and his team only responded to questions they thought made sense, especially from a technology standpoint. The aftermath of this approach was a complete disaster. Several company systems were heavily infected with malware, and the company intranet had to be taken down.

MANAGING INHERENT TECHNICAL ISOLATION (ITI)

In another related case study, *Ivey Business Journal* reported in 2002 that one of the biggest challenges facing IT departments was employees' buy-in and fostering an open work environment. The authors summarized this challenge as follows:

> It is important to develop a framework for the technology that the organization will use. The framework should take into account the nature of the work that individuals are expected to do and assign appropriate technology configurations that will enable employees to meet their business goals. This framework should also maintain technology within a particular work group at comparable levels, thus reducing intra-departmental disparities while making inter-departmental disparities more manageable.

In the analysis above, the authors identified that a non-inclusive technology culture was on the rise, and this culture clearly provided a level of operational and management disparities that needed to be managed. That same year, Mr. John Wash, the CEO of a small technology firm, was asked about this issue, and he gave what I would refer to as a blunter response. He said it is clear that most IT managers today are practicing or implementing technology systems and tools in a way that clearly demonstrates professional bias. He said he wondered whether IT managers thought that all employees within organizations graduated from colleges with some technology training.

My focus is on the impartial competency baseline approach to technology implementation. If you take the story of Van Lines Inc., you can tell how most IT leaders react to challenges coming from nontechnical versus technically savvy employees. This inherent bias, which I describe as inherent technical isolation, begins to grow at a very rapid level as long as it is not addressed by senior management. Interestingly, most companies never identify this culture until it is too late. Very often, some nontechnical staff will engage in a war of words with IT staff to get simple problems solved. For most IT managers, the culture of inherent technical isolation begins to attract their attention when team members are often failing to reach any effective communication resolution while talking with or trying to address nontechnical employees' technology-related challenges.

While it is difficult for most organizations or IT leaders in particular to admit to this challenge, many employees are often open to discussing these issues. In fact, there are three key indicators that often express, perhaps implicitly, the existence of such a culture. These indicators include (1) the IT department prides itself on sophistication over simplicity, (2) emphasis is often placed on tools and systems instead of people and collaboration, and (3) employees hate IT-related communication. Whenever you see these signs, IT leaders, senior management, and employees in general must begin to immediately work together to address these challenges. There are basic steps that can also be consistently practiced in curbing these challenges, including discouraging IT-related preferential treatment, building and empowering your workforce around technology goals, and promoting strategic innovation on technology development issues.

Without compromising the need to advance IT operations and improve service delivery, organizations can work to efficiently eradicate the culture of inherent technical isolation by consistently focusing on the three key objectives of information system security and/or technology needs. These objectives are confidentiality, integrity, and availability. To meet these objectives, IT leaders must work with the entire workforce to promote innovative thinking around technical collaboration that simply supports the impartial deployment or response to technology-related needs across the organization. Put simply, we must exercise professionalism in technology and information security without perpetrating professional bias. Essentially, IT leaders must simply continue to effectively emphasize the goal of information security without creating unnecessary administrative bureaucracy for employees, especially those with no technical background. Confidentiality, integrity, and availability are the key goals of an information system. Once these are fully communicated and well understood, employees would be able to play a vital part in

MANAGING INHERENT TECHNICAL ISOLATION (ITI)

leading efforts to promote strategic innovation regarding technical processes in the organization. Below we summarize the goals of an information security system.

- Confidentiality—In the realm of information security management, confidentiality refers to the protection of information from unauthorized individuals/organizations and processes. It involves preserving authorized restrictions on access to information, including means for protecting personal privacy and patented information. For example, when utilizing an online banking service, confidentiality requires that all customers be entitled to authorized access only. Nobody wants their private banking details to be compromised. Therefore, it is the responsibility of the bank, through its information system, to guarantee confidentiality at all times. A loss of confidentiality is the unauthorized disclosure of information.
- Integrity—When we talk about integrity, we refer to a process of promoting high moral standards or strong ethical values in the use of technology services. In the case of information security, integrity refers to the protection of information systems and processes from unauthorized modification or destruction. It simply means that organizations must have effective and efficient systems in place that safeguard the integrity of critical data. For this requirement to be met, both data and system integrity must be protected. Data integrity requires that any change in information services and processes be carried out in a specified and authorized manner only. System integrity, on the other hand, requires that a system acts in accordance with specified standards free from any or all forms of intentional unauthorized manipulation. In their book on computer information security, Stallings and Brown provide a good perspective regarding the breach of integrity as it relates to information security. The authors explained that, for example, when a nurse who is authorized to view patient records

deliberately falsifies a patient's allergy information outside of a formally approved process, such action amounts to a clear breach of integrity. Accordingly, a loss of integrity is the illegal alteration or destruction of information.

- Availability—While this requirement is usually overlooked, its key objective is by no means less important than that of confidentiality and integrity. In order to legally preserve and safeguard the integrity of data, the data must first and foremost be available. Thus, availability pertains to the assurance that a system is available, reliable, and open for use by authorized individuals at all times. For example, imagine going to the Division of Motor Vehicle (DMV) to renew your license plate but quickly realizing that the DMV's processing machine was down. This automatically impacts your ability to move forward with your vehicle's registration process. Incidentally, the delay thereof highlights the critical importance of availability as it relates to information systems usage. The same example can be applied to a host of other important information system services, especially those that are used in the service industry. Therefore, a loss of availability is the disruption of access to or use of an information system. Given the crucial nature of such a requirement, it is therefore important that organizations work tirelessly to ensure timely and reliable access to information systems at all times.

Addressing the digital divide

Most people agree that the central focus of governments around the world is to protect and cater to the security and well-being of their citizens. Nevertheless, implementing such a task comes with critical challenges. But in spite of the difficulties, many people continue to believe that experienced leadership is the key to achieving such an ambitious goal. However, one of the major challenges facing governments, especially those in developing nations, is the digital divide. Since the industrial revolution, many developed nations have experienced unprecedented growth and development, be it in health care, education, agriculture, technology, and more. Advanced tech products have helped to spur economic growth, reduce labor inefficiencies, increase economic opportunities, and improve the standard of living in many developed nations. Unfortunately, the reverse holds true for many developing and underdeveloped nations. Addressing this unfortunate situation is the major task facing current and future development professionals in both the public and private sectors respectively.

The digital divide is the terminology that describes the gap between those who do and those who do not have access to modern communications technology. It is a complex subject that deals with multidimensional issues including social, economic, and political. The digital divide presents several critical concerns including access to quality education, crime management, access to the Internet, access to affordable and quality health care, economic inequality, and the lack of the necessary IT infrastructure to support modern communications technology, particularly among disadvantaged social groups.

Despite the popularity of ethics, these concerns continue to exist. Today, in both developed and underdeveloped nations, millions of people lack access to quality education, and millions more do not have access to health care. In 2014, reports from the Federal Communications Commission (FCC) concluded that more than thirty million US homes lacked high-speed Internet. In 2010, as means of addressing this concern, former US President Barack Obama released a national broadband plan with the hope of giving "every American access to high-speed internet" by 2020. While there has been enormous progress, as of 2020, some twenty million Americans still lack high-speed Internet services. As opposed to the general myth that the digital divide is only a developing world problem, these statistics speak for themselves.

Addressing the digital divide remains a great moral issue of our time. Many of those who lack access to the Internet and modern communications technology cannot benefit from revolutionary IT services, including online information about new medical discoveries, crime management, investment opportunities, and many more.

Formal US President Bill Clinton eloquently addressed the moral challenge of the digital divide when he said: "It is dangerously destabilizing to have half of the world on the cutting edge of technology while the other half struggles on the bare edge of survival." Hopefully, innovative efforts like the E-Rate program (which provides money to connect schools and libraries to the Internet) as well as the One Laptop per Child (OLPC) initiative will do more to bridge the gap in the US and around the world. However, this effort can simply not be achieved without critical education and business innovation.

The role of education in addressing the digital divide

Education is indeed a solution in addressing the digital divide. The sharing of relevant information, transfer of knowledge through skills training programs, and the expansion of access to technology are major educational strategies needed to address the digital divide. Quality information leads to empowerment, and empowerment inspires motivation! If we effectively utilize the power of education and empowerment, we are more than capable of solving this problem. Essentially, the importance of education in addressing the digital divide cannot be overemphasized. For example, the majority of the people who lack access to modern communications technology simply do not have the skill sets to operate basic tech products such as laptops, iPads, and tablets. Strong efforts must be made to educate and train these individuals. Incidentally, shared education and experience can play a critical role in this process, especially through the creation of programs that focus on the expansion of access to modern communications technology everywhere. Expansion as a policy is key since many people still lack access to modern tech products and services. Meanwhile, creative thinking, advanced studies, and professional collaboration in the areas of hands-on training and knowledge transfer are methods that can also be used to not only bridge the digital divide but also spur new innovation for human development.

Many researchers, businesses, and development organizations have proffered critical solutions to address the digital divide. However, what seems to be lacking is the creative commitment toward a sustainable solution. While it is not yet clear how this will be addressed, I hold a strong belief that innovative education will be key to finding the big solution. To this end, this book recommends two practical solutions that, if effectively utilized, could certainly have a big impact in bridging the digital divide.

First, *information technology literacy* is a must. Global Internet World Statistics have suggested times without number that this is perhaps the biggest challenge facing policy experts today in the area of bringing technology to the needy. The group argued that due to the lack of basic computer skills predominantly among disadvantaged social groups, they are often excluded from many IT investment opportunities. Individuals and organizations often exclude these groups since it will cost time and money to train them. Moreover, businesses are always looking to minimize cost and maximize profits. On the contrary, bridging the digital divide will require some level of unconditional social inclusion built on a strong principle of moral intelligence. Educational programs must be developed and implemented so that majority of the people who lack basic technology skills sets can be fully trained. This could certainly help to bridge the gap we are facing now.

Next, we must aggressively expand access to *information communications technologies* (ICTs). One of the reasons why the digital divide exists is that many people still lack access to ICTs. Consequently, a dual approach to the situation requires both the transfer of technical knowledge through educational training programs as well as the sustainable expansion of access to modern communications technology. As mentioned earlier, even in advanced countries, several people still lack access to high-speed Internet. Moving forward, national leaders and policy makers must take into serious consideration the impact of information technology services on improving the quality of life. Hence, greater efforts must be made to ensure that people all over the world have access to high-speed Internet and information communication technologies without any hindrance.

We must remember that throughout history, the power of education has been countlessly tested. Coincidentally, for those who believe in education, every critical test has only pushed humanity to continue to seek a better understanding of the actual value of education to mankind. Today, the digital divide is one such test. This challenge is pushing humankind to employ every resource available, including education, technology, and human capital, in an effort to address the problem. Personally, I believe that the solution is here. The answer for me lies in responsible production, target-oriented education, and sustainable expansion of access to information communications technologies everywhere. If we get the balance right, once again, the power of education will have subjugated the fear of doubts. Humanity may have been and still remains a victim of many things including war, hunger, poverty, and diseases, but I strongly believe that we now have the capability to end the digital divide. Yes, innovative education is the solution!

CHAPTER 5

Enhancing Leadership through Efficient Technology Solutions Adoption

The final chapter of this book places emphasis on how technical applications can support enterprise improvement across both internal and external business processes. Chapters 1 through 4 emphasized the role of experienced leadership in supporting or driving technology goals. This chapter will focus on how to deploy useful technology systems using a very important business intelligence approach. There are so many business applications today that are important for business development. However, I have chosen to focus on one of the breakthrough technology applications that I believe has truly disrupted and continues to positively change business operations across different industries—the enterprise resource planning (ERP) system. In this chapter, we will provide an overview of the system, cross-examine its implementation, share important insights for evaluating ERP vendors, and finally, end with explaining a few simple steps to deploying an ERP system.

An enterprise resource planning (ERP) system is a business concept that focuses on integrating the functional areas within a business. ERP concept emerged as a result of advancements made in computer hardware and software development in the 1960s and 1980s. For years, the functional business model had dominated business operations. In the functional model, planning, organizing, and management of activities were mainly carried out in line with the functional responsibilities as specified within an organization. The functional model works in a way that allows each business unit to collect its own data and make decisions as needed. Such a process was, however, carried out in a disintegrated business environment. Eventually, as the demand for efficiency in business operations became crucial, the need to eliminate information silo grew paramount, and companies could no longer afford the level of ineffectiveness offered by the functional business model.

With an effective ERP system in place, business managers are able to focus on collecting important data that can enhance business efficiency, improve service delivery, and also help to maximize profit. ERP systems stand out particularly for their key objective in driving efficiency across business processes through promoting effective communication and greater collaboration across the functional areas of the business. Moreover, given the growing demand for big data in modern business, the importance of implementing an ERP system cannot be overemphasized. ERP systems not only work to centralize data in the organization but to also improve the quality of data collection and analysis. Some studies have also shown that when a business effectively utilizes an ERP system, it experiences improved collaboration, better data analytics, improved efficiency, greater control over security, improved inventory and production management, and better customer relations management. Other benefits that come from adopting an ERP system include reduced IT-related training costs and a more synchronized strategic management planning. But while ERP systems offer enormous benefits, especially with regard to business process improvement, there are also challenges with the system deployment, usage, and management. Using a hypothetical company, we will review some of the pros and cons of deploying an ERP system.

The positives and negatives of deploying ERP systems

Bunny Steel, located in Denver, Colorado, functions by supplying steel to organizations and individuals working in the construction industry. The functional areas of the business include supply chain management, human resource, finance and accounting, and sales and marketing. For this case analysis, this book assumes that the company's information system is currently not integrated and, to a large extent, operates the traditional functional business model. To date, there is not much visibility of cross-functional teams working to promote collaboration that can help alleviate the challenges presented by the disparate information system. However, the company has experienced considerable successes over its long years in business operations and has managed to remain competitive in spite of the modern business challenges. Starting from a humble beginning as a small family business in the 1950s, today the company has expanded to almost every state across the United States and now sells nearly five hundred thousand tons of steel per year.

Notwithstanding, Bunny Steel continues to face the challenge of information inefficiencies which has resulted in the lack of effectiveness across the business management processes. This challenge exists for several reasons, but a key among them is the lack of a fully functional integrated information system. Therefore, having carefully examined the key functional areas of the business as well as the historical and cultural practices promoted by its management team, it is logical to conclude that implementing an ERP system could have both positives and negatives for the company. First, the company will benefit from an integrated information system in the following ways: improved communication and collaboration across the business platform, improved efficiency across the business processes, easier global expansion, reduced information technology (IT)-related costs, greater management control over information, and improved security and data quality.

Integrated information systems work best to reduce information disparities and improve operational efficiency. When a company's information system is fully integrated, it improves communication due to the increasing frequency of cross-functional teams working to promote effective process management as opposed to focusing on the functional areas alone. Essentially, an integrated information system promotes centralized data management which affords management the opportunity to have greater control over the business processes and also enables effective monitoring of key activities as well as the development of standards for overall process improvement. Additionally, a company benefits from an integrated information system in areas such as IT-related costs reduction and faster service expansion. Many IT professionals have also argued that other important benefits of an ERP system include its ability to bridge the barriers of currency exchange rates, language, and culture, thus making it easier for companies to expand at the global level. Moreover, since ERP system integrates people and systems into a single information platform, it reduces IT training costs by allowing management to focus on a single comprehensive training rather than spending company resources to train people on the usage of different systems across the business functional areas.

Despite these potential benefits, the implementation of an ERP system at Bunny Steel could also bring about some key business setbacks. Let us look at some of the possible challenges that could emerge during or after an ERP system deployment at the company. Given that ERP systems are very expensive, each company has to carefully study its financial position before deciding to adopt an ERP system. However, whether or not a company decides to use ERP, the decision should be based on a sound business analysis rather than the traditional feeling of resistance to change. For the relevant company, some of the potential negatives for ERP implementation, among other challenges, could include cost escalation, change in business culture, data migration challenges, and vendor lock-in.

Also, with over five hundred employees at the company, implementing ERP presents a major financial challenge. The Technology Evaluation Centers reported in 2013 that companies with one thousand employees or more are likely to spend between $100 to $500 million for an ERP system with operations involving different languages, currencies, and tax laws. Furthermore, there are other fees such as software installation, license fee, and consulting fee. Another challenge is that ERP, by design, implies system change and therefore requires users to adopt by making several institutional changes. This comes into play most often when a particular business, for example,

MANAGING INHERENT TECHNICAL ISOLATION (ITI)

sees some of its legacy systems as key and may only want to pursue partial integration of the ERP system. Such a process can be very challenging since different software will have to be used in dealing with what is sometimes referred to as configuration crisis.

Furthermore, data migration could present another major setback. Migration of disintegrated data to a new ERP system can be difficult and could sometimes go on for years. Consequently, achieving full data migration, which enables the smooth running of the ERP system, consumes time and often leads to intensive planning incorporating greater use of the company finances. Also, the risk of vendor lock-in as a result of engaging in an ERP contract without carefully evaluating the vendor's proprietary rights as well as their product flexibility could be one of the most dangerous challenges. Therefore, when deploying an ERP system, management must engage in a detailed planning process to fully examine the business functions and processes, personnel capabilities, and financial stability before deciding whether or not to adopt an ERP system.

Leading organizational ERP assessment project

According to this case study, while the company information system may be non-integrated, its functional areas, including sales and marketing, supply chain administration, finance and accounting, and human resources, are very much interdependent. Sales and marketing plan for product promotion, collect sales data, generate sales order, and work to continuously market the company products. Nevertheless, the sales and marketing department needs information from all the other areas to carry out its functions. Similarly, the supply chain administration section uses sales records to plan product scheduling, update inventory, and set up accurate shipping and delivery dates for customers. Moving forward, data from both sales and marketing and the supply chain management are needed by the finance and accounting section for budgeting purposes, customer credit financing, and to also maintain updated accounting records. Also, the human resources section uses sales forecasts information, accounts receivable records from customers, and production planning records to determine staffing and personnel needs.

Therefore, these functional areas, while operating independently, also rely on each other for the purposes of completing the overall business process. In today's business environment, we must remember that some customers are not really concerned about the individual departments within a business function but are more focused on the level of effectiveness applied by each individual unit in the delivery of effective customer services. To this point, utilizing an integrated information system presents a great advantage for businesses. However, like most major development projects, ERP implementation has its pros and cons. For this company, the pros and cons for each of the functional areas highlighted above are as follows. For the *sales and marketing department*, the pros for ERP implementation include improved customer relations management and complete synchronization of sales and marketing activities, while the cons could be setbacks in customer engagement due to change in business approach and delay in sales data migration to the new system. For *supply chain management*, the pros include improved inventory record management, integrated shipping, and effective delivery service solution. Meanwhile, the cons could include increased costs for new software and hardware as well as the length of time it will take to complete a detailed training for targeted personnel on how to use the new system.

Moving forward, the pros for the *human resources department* could include but are not limited to reduction in the use of paperwork due to full automation of the system and improved data collection and human resource reporting as well as employee self-service options. Nevertheless, the high cost of system integration and differences in departmental information collection priorities could present negative feedback for an integrated human resource system, especially when it comes to adopting a centralized data management approach. Finally, the pros for the *finance and accounting* section would be improved efficiency in collecting and reporting accurate financial information and improved data quality and better financial forecasting. However, the cons could include high cost of acquiring the software and hardware as well as challenges to effective system configurations needed to meet the business's current financial standards. Paying attention to these pros and cons helps put organizations in a knowledgeable position to lead adequate preparation for an ERP system deployment and management.

At this point, we now have some basic insight into what an ERP system is and what value it can provide organizations, especially if all the major challenges that come with deploying the system are effectively handled. In summary, I love to describe ERP systems as a set of technology solutions that really help you visualize your organization or business more like a process as opposed to the traditional understanding of simply looking at your organization as a group of individual units implementing different or specialized functions. ERP systems help business leaders to not only see but to clearly understand functional interdependence within the organization. It improves collaboration by helping to identify communication baseline, define communication goals, and implement communication and collaboration policies. There are several ERP systems out there that can be adopted by organizations. This book will review two of the most competitive solutions in the supply chain area.

ERP supply chain management solution review

Today, the increasing demand for goods and services coupled with the idea to improve customer service experience is forcing organizations to pursue new and innovative paths to meet these growing challenges. Essentially, meeting these challenges also helps competitive organizations to keep up with the market competition. To this end, organizations looking to improve customer relations, reduce production inefficiencies, and maintain accurate inventory levels, among other things, are now pursuing the concept of supply chain management (SCM) enterprise resource planning as an essential solution. In the below paragraphs, I have provided a quick high-level overview of a few supply chain management ERP solutions.

Supply chain management deals with all the interrelated business processes which occur within an organization from product sourcing to packaging and shipping of finished products (NC State University 2017), while enterprise resource planning, on the other hand, focuses on the integration of the functional business processes—that is, human resource information system (HRIS), supply chain management (SCM), customer relations management (CRM)—and financials into one unified information system in order to streamline processes and promote overall efficiency in business service delivery.

With the multiplicity of supply chain management solutions available today, organizations are faced with the difficulty of making decision, especially when it comes to selecting a supply chain management software. Basically, implementing this process usually requires an intensive management planning. In most organizations, the planning process requires business managers to focus more on the business process with the goal of delivering excellence in customer service through improved decision-making. Meanwhile, best practice suggests that the planning process should not deviate from the central reason why organizations seek the supply chain management solution. Some of the potential reasons why organizations would seek supply chain ERP solutions may include reduced overhead and operational costs, improved efficiency across supply chain processes, decreased operational errors, automated workflow, and flexibility in meeting changing markets demands. These benefits, however, must come with intensive cross-functional planning and substantial human resource inputs to the system. Thus, one thing that should be remembered is that in relation to the overall business process, supply chain management is never considered a stand-alone concept.

In the next few paragraphs, we will compare and contrast the supply chain management solutions offered by two different vendors. Namely, Oracle, and systems, applications, and products in data processing (SAP). Additionally, I will also examine some of the required IT products, including hardware, software, and databases, needed to facilitate the implementation of a supply chain management ERP system within an organization. The major goal of this section is to provide the reader with a foundational understanding of how individuals, businesses, and organizations can select a supply chain management vendor/solution for their respective organizations.

Similarities and differences between SAP and Oracle SCM products

Before I examine the ERP products' similarities and differences, I will first provide a quick overview of each of the products. Interestingly, both SAP and Oracle offer integrated business planning software as a supply chain management solution. SAP integrated business planning focuses on demand management and profitability analysis by effectively utilizing real-time supply chain planning data. Designed with a very powerful in-memory computing technology, the software efficiently combines sales and demand analysis, operations and supply chain planning, and focuses on inventory optimization.

On the other hand, Oracle integrated business planning supply chain management solution provides a strategic planning advantage that helps businesses align financial and operations objectives while applying best practices across the business life cycle to meet corporate profitability and other operational goals. The software helps organizations to develop quick consensus regarding strategic and operational goals and uses key performance monitoring metrics to deliver real-time results that help support effective management decision-making. But what are the key similarities and differences between these two products? The table below depicts some key similarities.

Comparing SAP and Oracle SCM products

Category	SAP IBP	Oracle IBP
Product information	Cloud-based solution offering scalability, flexibility, and cost efficiency	Cloud-based solution offering scalability, flexibility, and cost efficiency

Integration functions	Integrates sales and operations planning with other functions across the SCM cycle	Integrates sales and operations planning as a major function in the production planning process
Product focus	Designed for small, medium, and large businesses	Designed for all sizes and types of organizations
Lead feature	Integrated supply chain management tool for effective operations planning	Built-in supply chain management system for effective order management and SCM operations

Notwithstanding, both vendors do offer distinct technical features in their various SCM solutions. To effectively examine the products' dissimilarities, I will provide a brief description of their key features. SAP integrated business planning has several key capabilities including supply chain control tower, forecasting and demand management, inventory optimization, sales and operations planning, and response and supply planning. Each of the above features is built with a unique capability tailored toward improving overall efficiency in supply chain management processes.

The Oracle integrated business planning, on the other hand, has the following key features: configurable process-definition templates, embedded analytics, consensus forecasting, best practices in sales, and operations layout and tables. Other key capabilities include a supply chain management control solution as well as a customizable data analytics model which supports customers' data migration and demand management services. The table below depicts some of the major differences between the two products.

MANAGING INHERENT TECHNICAL ISOLATION (ITI)

Contrasting SAP and Oracle SCM products

Features	SAP IBP	Oracle IBP
Supply chain planning	SAP supply chain planning improves supply-chain-related performance by consistently monitoring and measuring supply chain alerts and metrics in real time.	Emphasizes consensus planning using different collaborative planning capabilities. The feature combines different marketing, sales, and operations forecasts utilizing configurable weightings.
Sales and operations planning	Aligns sales and operations planning with the business strategic goals. Increases profitability via increased market share	Focuses on improving demand management as a means of increasing productivity which subsequently leads to greater revenue generation
Distribution management	SAP has a more sophisticated event management capability which gives businesses better insight into their entire supply chain management cycle.	Has a more advanced planning system (APS) allowing you to track cost and efficiently allocate resources

| Integration support | Integrates global customer relations management (CRM) with SCM functions at a rapid level | Focuses on integrating specific SCM operations in line with service needs and various business demands |

Deploying these solutions requires critical due diligence, especially in the area of information technology infrastructure needs assessment. Therefore, with regard to the IT product requirements, including software and hardware that are needed to utilize both Oracle and SAP supply chain management solutions, the onboarding processes executed by both companies have clear differences. SAP integrated business planning software is deployed through a three-step approach. First, customers go through the introduction to the integrated business planning system and applications. Next, SAP uses a cloud platform integration (CPI) link to enable on-premises integration and set up user access for system support services. Finally, the third stage is data integration. Using the standard cloud platform integration data services, SAP integrates the cloud software with the organization's on-premise business system.

Similarly, the complete software implementation process for Oracle is carried out through the following steps: planning, configuration, setup, deployment, and maintenance. At each stage, a specific technical task is executed. During the planning stage, the company determines the offerings that a business may want to implement and evaluates the product features against the company SCM processes. Next, at the configuration level, system implementers opt into the offerings by matching functional areas and features that best fit the business requirements. A key thing to remember at this level is that your IT team should be able to determine product configuration needs in alignment with the software configuration capability. For the setup stage, the major task revolves around data entry and application testing. This process can be done over a period of time to identify system bottlenecks before the software is officially deployed. Once the setup stage is over, verified data are then moved from the testing environment to the production environment, thus initiating full deployment. The next and final stage is ongoing maintenance and system support. Essentially, as the business requirements change over time, adjustments are made to maintain the product while adjusting its capabilities to meet new business demands.

Evaluating ERP systems implementation success

It is important to note that whether the ERP solution deployment is successful or not is totally dependent on the project deliverables defined in the statement of work. For most organizations, a successful project implementation is characterized by the efficient and timely accomplishment of the objectives specified in the project requirements. However, to ensure project implementation success, project managers must design the project framework and effectively coordinate both human and material resources to meet the project objectives. The need for effective coordination and collaboration cannot be overemphasized. This need indisputably involves the utilization of various project management techniques along with a capable leadership team to deliver results as planned. Basically, a successful project implementation requires the management of important issues including project integration, communication challenges, cost management, scope validation, and risk management. Achieving these goals often requires critical organization planning.

In several organizations, effective planning is used to support change management and project success. Without proper planning, it is almost impossible to determine the outcome of any project. Therefore, enterprise resource planning is a critical business initiative that must not only be carried out with the proper plan but must also be implemented with the aid of an experienced team. Whenever an organization sees it fit to implement an ERP project, using an experienced team is always important. Without an experienced team, it is almost impossible to manage any ERP deployment project. A key example we want to spotlight is the massive failure of ERP projects a few decades ago. By the end of the 1990s, the exponential growth in the demand for ERP solutions was met by a major shortage of experienced consultants. Many organizations that needed ERP solutions could not find individuals with the requisite skills to lead successful ERP project implementation. This condition eventually led to the failure of most ERP projects during that period.

Despite the complexities associated with ERP projects' implementation, studies have shown that with the right plan in place, backed by a supportive management team and well-trained employees, ERP projects can be successful. Nevertheless, it is important to also mention that no matter the level of organizational preparation, some unexpected challenges could still emerge along the course of an ERP project implementation. Therefore, it is important for organizations to prepare well in advance by developing a comprehensive implementation framework that is not only used to deploy the ERP solution but is also flexible to change, especially when it comes to meeting the needs of changing business environment. There are steps that can also be taken to help organizations evaluate ERP solutions and vendors.

To this end, I will focus the next few paragraphs of this chapter on addressing some important steps that organizations can use to enhance their evaluation approach during an ERP project implementation. The first part of the discussion will focus on some of the essential steps and activities that individuals and organizations can follow to ensure successful ERP systems implementation. This part will include a clear approach on how to lead a successful evaluation of ERP vendors and their respective software solutions. The second part will address a proposed framework for implementing an ERP system into a full enterprise production system. I will then conclude the chapter with a set of operational measures that can be used by organizations and project managers to determine if an ERP project was successfully executed or not.

Conducting enterprise ERP needs assessment

Like most strategic business operations, when seeking to implement an ERP solution, organizations should begin with the business's primary needs assessment. To this end, critical planning along with the availability of trained personnel are two key essentials at the center of an ERP project implementation. Notwithstanding, the level of technical details needed for an ERP project to be successful over a period of time goes well beyond the plan and the team. Over the years, several failed projects across many organizations have justified how critical ERP implementation experienced is for project success. Essentially, organizations cannot simply leave this process to external consultants or prospective vendors. To conduct a value-added needs assessment, some basic steps must be followed. And while these steps may differ across project teams and organizations, the underlining objectives are very similar. To explain this approach, this text will focus more on the manufacturing industry.

Companies that are working in the manufacturing industries can improve efficiency across the enterprise by following the below-proposed approach to conducting an ERP needs assessment. First, the process begins with an intensive project preparation. Next, strategic emphasis is placed on the development of a clear business case for action. After developing the business case, the project management team should then conduct a robust market survey to access possible solutions that are in alignment with the business' required objectives. Finally, the project team should execute test demonstrations and review application support needs before officially moving forward with any selected ERP solution. The steps are explained below.

At the *project preparation stage*, the company must set up clear personnel qualification requirements for selecting the project team. The project team is then authorized to define the project scope, determine the hardware and software needed to support the ERP system, and also develop a budget estimate for the project. Next is the *business case stage*. When developed, this stage should provide documentation that clearly communicates the project requirements and must clearly explain how the project requirements align with the company's strategic vision. Notwithstanding, the business case should also specify the functional processes that will be included within the ERP system and how they will impact overall process improvement. Documents produced at this stage are very important since they can also be used in guiding consultants along the final stages of the project implementation.

After the above steps have been taken, the organization should then use the business case and all other relevant documentations to set up standards for a *market survey*. During this stage, a business should make all effort to utilize as many sources as possible including the internet evaluation center, industry trade journals, personal references by friends and colleagues, as well as recommendations from consultants. The market survey should, however, have a clear objective. All of the short-listed vendors emerging from the market survey must meet those objectives as specified within the company business case. After selecting the initial vendors, the project team should verify via an evaluation exercise the suitability of each vendor to the company's ERP needs. The goal would be for the competing vendors to clearly prove how each of their ERP solutions aligns with the business' project requirements as well as the organization's need(s) for any future extended support service(s). When the initial demonstrations are concluded, the number of vendors should be narrowed down to at least two or no more than three as the project team prepares for next steps. The next stage is the final selection. Before moving to the final phase, the project team should put out a final request for information (RFI) for the remaining vendors, requesting each of them to detail out their most realistic budgets as well as any final modification to their original ERP implementation plans.

When a vendor is finally chosen by the company, the final step is to execute *test demonstration* and *live support*. During this stage, the project team usually begins using the ERP software, and a help desk is set up to enable system support from both the vendors as well as project consultants if available. The organization leadership, via its project management team, should decide how long this process will last and must try to select a final project closing date as soon as possible to help control cost. At the testing stage, as challenges emerge, all necessary adjustments should be made to account for changing business needs. Nevertheless, the ERP system itself should always be customizable, especially to support adaptation and operational improvement over time. This is very important because as users continue to interact with the system, operational challenges will be identified which might have not been addressed during the official ERP project implementation circle. Put simply, companies should view ERP projects as an ongoing strategic development process rather than a destination.

To further enhance enterprise productivity, business leaders across the manufacturing industry can implement an ERP system to help improve specific functional business processes. A successful ERP project can improve efficiency across the following functional areas including supply chain management, human resource, production, and sales and marketing divisions, respectively. For example, if a company opts to use systems, applications, and products in data processing (SAP) ERP, functional business processes such as production planning, material requirement planning (MRP), production scheduling, inventory management, and shipping will have to be carefully analyzed and fully structured to help maximize the benefits of the ERP software. Other functional business processes that would also need to be carefully evaluated for potential ERP system benefits include promotions planning, sales order management, accounting and invoicing, staffing, and standard cost versus actual cost of production analysis. Once the internal assessments are completed, organizations can now take the below steps to evaluate ERP service providers.

Evaluating ERP vendors and their products

Evaluation as a concept involves the analysis/examination of systems, procedures, practices, products, or individuals to help determine or to match capabilities against desired objectives. Simply put, the purpose of an evaluation is to inspect readiness level and use the data to inform any particular decision-making process. Therefore, in line with the overall objective of implementing an ERP system, the following steps can be used to carefully evaluate ERP vendors and their respective products:

- A. Determine your project requirements.
- B. Develop personnel qualification and experience requirements.

C. Establish your vendor evaluation criteria.
D. Develop the budget and implementation timeline.
E. Select a short list of vendors.
F. Compare potential ERP solutions against the project requirements.

Step one: determine the project requirements. As indicated earlier, the key business processes that most manufacturing companies would want to improve via an ERP system implementation include production planning, material requirement planning (MRP), production scheduling, inventory management, and shipping. Others are sales forecasting, promotions planning, sales order management, accounting, staffing, and standard cost versus the actual cost of production estimation.

To begin the process of evaluating ERP vendors, organizations must first develop a list of clear requirements for the project. Usually, such a list is developed as part of the organization's business case. It is very important to keep in mind that the business case is the guiding document that lays out the strategic vision of the organization and clearly explains its strengths and weaknesses as well as the critical need for the ERP solution. Also, the same document must clearly communicate the ERP project objectives, potential vendors evaluation strategy, and, most importantly, the business case must detail the current challenges impacting functional business processes within the organization. Understanding the importance of the project requirements and subsequently developing them is the first step that organizations must follow in order to begin the process of evaluating ERP vendors. Essentially, all needs are developed at this stage and the project team works to establish what leadership and automated solutions are required to address these needs.

Step two: personnel qualification and experience. Project development and implementation is a technical endeavor. Not all employees can boast of the requisite skills needed to ensure successful project implementation. Moreover, while most ERP project implementation may be done via certain conventional standards, a number of the fundamental pattern for developing and implementing a project including details such as technical requirements, business competency baseline, resources management, and technological complexity often differ across projects. Therefore, at this level, organization leaders must carefully evaluate the project scope and gather sufficient technical information in order to set up qualification and experience criteria for each potential team member. Additionally, this is very important for two major reasons. First, without a competent team, a project is most likely to fail. Secondly, even if you select the most qualified team in the world, they will still have to go through additional and/or specific management training which is sometimes very rigorous depending on the project complexity. Thus, defining the qualification and experience criteria is vital to reducing any knowledge gap and ensuring implementation success. It is therefore advisable that this step be taken very seriously and pursued with the greatest possible caution. Always remember that it is the project team that is going to evaluate your potential ERP vendors. Therefore, the qualification and preparation needed for the team cannot be compromised.

Step three: establish vendor's evaluation criteria. Once the project requirements have been determined and the leadership team has evaluated and chosen the project team, the next stage is to develop the official evaluation criteria. This is probably the most sensitive stage in the evaluation project implementation process. This stage is very sensitive because it incorporates two technical dimensions. First, it involves the review, simplification, and clarification of internal business needs. And next, it also requires an extensive baseline assessment of potential vendors' products as well as their holistic capacity for the ERP system deployment. From my experience, I would even suggest that when developing the evaluation criteria, organizations must pay key attention to technical issues such as deployment cost, scalability, vendor's ERP implementation history, integration support, product usability, customization functions, training, and support needs. As I mentioned earlier, it will take a competent project team to document these technical criteria and compare them very carefully against the business needs or the desired ERP solution. Finally, this stage should produce the *official evaluation tool* that the organization will use to inform the next course of events in the evaluation process.

Step four: develop the budget and implementation timeline. Producing a project budget is not simply a traditional project requirement. Like most business management control standards, the project budget is required for multiple reasons. A 2015 project implementation research at Leading Change Inc. highlighted three important reasons for developing a budget. First, the organization argued that budgets are developed to determine a clear or holistic financial implementation road map for any undertaking. Next, they also claimed that a budget is developed to set transparency standards and ensure accountability before, during, and after a project implementation. Finally, the organization stated that budgets are developed to control project implementation cost and facilitate effective decision-making during project implementation. These reasons, coupled with other reasons such as trade, value exchange, and other traditional purposes together, justify why it is very important to develop a project budget for any major enterprise undertaking.

Moreover, at this stage, the project team would now have all the information needed to move forward with the project. Using such information, the next step is to match financial resources with the project needs and come up with a comprehensive project budget. Also, at this stage, the project team uses different project estimating techniques to determine an estimated time frame for the official commencement and completion of the project. Once these administrative requirements have been satisfied, the project team can now move forward with sending out a request for information (RFI) and have interested vendors submit bids to meet the requirements.

Step five: select a short list of vendors. After receiving the required number of proposals from interested vendors, the organization should move to close the RFI process and begin reviewing the submitted proposals. This stage may require several different approaches based on what has been outlined in the business case. However, two key things must be implemented with adequate due diligence. First, the project team must use what they have—that is, the predefined project requirements as well as the official evaluation criteria to review the submitted proposals. This can be carried out based on an agreed-upon administrative review procedure. It can be done either via a group or through individual review process. Likewise, it can be done in person or virtually. Next, a specific number of vendors who meet a significant portion of the baseline requirements should be chosen and brought forward for interviews, inquiries, and other business-related investigations covering both the vendor's business history as well as their product capabilities. During this stage, live demonstrations must be carried out so as to determine the most competitive vendors in relation to the actual project needs.

Finally, this stage ends with the selection of short-listed vendors that will be duly informed about the next round of scrutiny. In most cases, the short-listed vendors might not have necessarily satisfied all the project requirements but would have met certain fundamental standards that align with some of the project critical goals at various levels. Based on this logic, the organization should seek to move forward with further examinations of the short-listed vendors in order to determine how well they can modify their solutions to fit the project's specific needs. It must be noted that very often across ERP implementation projects, no 100 percent solution is guaranteed. However, the organization, at this stage, should seek a solution that highly impacts a significant level or perhaps the majority of the project requirements. Also, from a cost perspective, vendors' proposals should be carefully evaluated against the official project budget, and where necessary, adjustments should be made to the budget once a specific vendor is chosen.

Step six: compare potential ERP solutions with the project requirements. Comparing short-listed vendors and their various solutions is a crucial stage in the evaluation process. At this point, the organization or project team is often equipped with sufficient information that can be used to easily determine its top two vendors for final selection. Additionally, this is where critical questions are asked, and the project team can make demands with a considerable level of confidence for more in-depth product knowledge from the remaining vendors. During this stage, a combination of test demonstrations, especially issues relating to deployment efficiency and effectiveness, product usability, integration support, scalability, data security, and configuration needs are all carefully examined. The project team initiates a careful review and evaluation of each of these technical requirements and uses the data from the live tests to help inform the final decision-making process. Users' feedback are also encouraged, and the determinations thereof are incorporated into the final vendor selection process.

Meanwhile, from the evaluation approach, one can clearly determine how planning is very important in project management and implementation. Along each stage, we provided a step-by-step guide demonstrating how an organization, through its project team, uses a set of predefined project requirements as well as other administrative tools at their disposal to clearly control the vendors' evaluation process. Moving forward, it is important to mention that without the availability of a solid plan and a well-experienced team, organizations desiring ERP solutions would almost be left to the mercy of persuasive marketing strategies from the vendors' market. Essentially, organizations that seek to implement ERP applications solutions must always bear in mind that unlike most business applications, ERP are critical because their impact can affect several areas of the business, especially data, operations management, communication, and risk management planning, respectively. Consequently, the six evaluation steps listed above are integral to the strategic management and operational processes that an organization can pursue when looking to successfully implement an enterprise resource project.

By mastering the ERP vendors' evaluation process, organizations can move to implement any ERP solution of choice with clear accountability. However, like other important management endeavors, it is vital that an organization's approach to implementing an ERP system begins with the business needs assessment. The project team must substantially examine the internal and external project requirements and carefully determine the organization's capabilities, including human and infrastructure resources. This is important because ERP systems are costly and can sometimes take years to implement, especially without the right baseline capacity assessment. It is also important for organizations and project managers to carefully analyze and match the potential capability of a prospective ERP solution with the actual needs of the business. Failure to do so could result in unwanted consequences that can have far-reaching effects on the business. For example, in 2017, when the management of one of the corporate manufacturing giants called Hershey Candies opted to pursue a holistic ERP system implementation, the company had failed to approach its business challenge from a practical perspective. Thus, after engaging systems, applications, and products (SAP) to implement a ten-million-dollar project, a series of critical management challenges emerged immediately. The resulting implications were chaotic as the entire business system broke down due to a poorly implemented ERP project (Pemeco 2015).

Eventually, not only did the company lose over $150 million in revenue but they also experienced a drastic reduction in share price by 19 percent and later lost 12 percent in their international market share. The recovery process was slow, and frustration increasingly grew around the business processes as they fought their way back up. Put simply, stories like these remind us why it is important to first understand the business needs and then approach an ERP system implementation with the greatest possible caution.

Implementing enterprise ERP solutions

Today there are a variety of ways to approach ERP implementation. Depending on the implementation team and the overall project management maturity level, the project can go from meeting simple standards to creating adjustments to meet complex business solutions. The relevant organization or project can determine what is best for them. And while there are multiple ways to approach the ERP implementation project, this book recommends the below approaches. Nevertheless, this text will only focus on explaining the first of the three methods.

1. Company-wide installation
2. Unit-by-unit installation
3. Key process installation (CYOP Consulting, n.d.)

As indicated above, the various implementation frameworks are available to provide organizations with choice elasticity when it comes to embarking on an ERP system implementation project. Each implementation approach has a unique objective. The objectives, along with the business value of the first approach, are carefully discussed in the succeeding paragraphs.

Company-wide installation. This is one of the most common approaches to ERP systems implementation. Through this framework, organizations implement a predefined step-by-step process to deliver a holistic enterprise-wide ERP system solution. This approach was very popular in the late 90s as many organizations looked to become Y2K compliant. According to the United States Securities and Exchange Commission, Y2K compliant brought about the development of software and hardware technologies to meet the demands of growing technical complexities across enterprise project management solutions, especially in the early 2000s. Many of the technology requirements at that time informed the departure from computer programming that previously stored the year value as two digits to the currently accepted millennium computer programming language. These new developments in computing began laying the foundation that eventually paved the way for the emergence of automated advanced ERP systems.

The major objective of the company-wide installation approach is to impact the entire organization by integrating the business functional processes into a centralized system based on a one-time approach. This approach is very challenging as the learning curve can be intensive and can sometimes even result in a greater need for high-level technical training and employee development. Also, given the complexities as well as the inherent changes that come with ERP system implementation, CYOP Consulting, a leading IT business firm, argued that this approach is most suitable for start-up organizations or small enterprises. The logic is that start-up organizations are in the process of developing, and as such, they are still exploring different enterprise practices in pursuit of their ideal management identity. Therefore, it is assumed that a holistic ERP implementation would definitely not disrupt but rather provide an opportunity for the business to explore best practices and can also add value to each identified business process. Notwithstanding, once an organization has developed a comprehensive ERP implementation project plan, any size or type of organization can decide to use the enterprise-wide implementation approach or the other two methods highlighted. We will begin *Managing ITI Part 2* by dedicating a whole new chapter to methods two and three.

ACKNOWLEDGMENTS

To my mom and dad, Mr. and Mrs. Daniel P. Charles Sr., thank you for your unwavering support to my education, passion, and continuous development. Your endless encouragement was an integral support system for completing this book. I have yet to see a belief as strong as the one you have always shown me. Thank you both.

To Dr. Rick Livingood, Messrs. Thomas Zordani, and Jeff Gross, I could not have asked for a better intellectual and moral support team. The passion and dedication you all brought to this work made each and every morning feel special. Your belief and support in this work will forever live on the sands of time in my heart! Thank you.

REFERENCES

Bassellier, G., and I. Banbasat. 2004. "Business Competence of Information Technology Professionals: Conceptual Development and Influence on IT-Business Partnerships." *MIS Quarterly* 28, no. 4: 673–694. https://doi.org/10.2307/25148659.

Danaher, M., K. Schoepp, & A. A. Kranov. 2019. "Effective Evaluation of the Non-Technical Skills in the Computing Discipline." *Journal of Information Technology Education* 18: 1–18.

Downing, C. G. 2013. "Essential Non-Technical Skills for Teaming.: *Journal for Engineering Education* 90, no. 1: 113–117. https://doi.org/10.1002/j.2168-9830.2001.tb00577.x.

Farrer, L. 2019. "Beware: Professional Isolation Is More than Loneliness." *Forbes*. Last modified February 15, 2019. https://www.forbes.com/sites/laurelfarrer/2019/02/15/beware-professional-isolation-is-more-than-loneliness/?sh=1364bc652723.

Freed, S. E. 2014. "Examination of Personality Characteristics among Cybersecurity and Information Technology Professionals." *UTC Scholar Archive*, https://scholar.utc.edu/theses/127/.

Golden, T. D., J. F. Veiga, and R. N. Dino. 2008. "The Impact of Professional Isolation on Teleworker Job Performance and Turnover Intentions: Does Time Spent Teleworking, Interacting Face-to-Face, or Having Access to Communication-Enhancing Technology Matter?" *Journal of Applied Psychology* 93, no. 6: 1412–1421. https://doi.org/10.1037/a0012722.

Livingood, R. A. 2003. "Predicting the Success of Potential Information Technology Professionals by Correlation to the

Myers-Briggs Type Indicator." PhD diss. Capella University, Minneapolis, Minnesota.

Lutes, K, A. Harriger, and J. Purdem. 2009. "Do Introverts Perform Better in Computer Programming Courses." *American Society for Engineering Education* (ASEE Conference and Exposition in Austin, Texas): 14.496.1–9.

Marshall, G. W., C. E. Michaels, and J. P. Mulki. 2007. "Workplace Isolation: Exploring the Construct and Its Measurement." *Psychology & Marketing* 24, no. 3: 195–223. http://doi.org/10.1002/mar.20158.

Misra, R. K., and K. Khurana. 2017. "Employability Skills among Information Technology Professionals: A Literature Review." *Procedia Computer Science* 122: 63–70. https://doi.org/10.1016/j.procs.2017.11.342.

Rasch, R. H., and H. L. Tosi. 1992. "Factors Affecting Software Developers' Performance: An Integrated Approach." *MIS Quarterly* 16, no. 3: 396–413. https://doi.org/10.2307/249535.

Shuter, J. 1984. "The Information Worker in Isolation: Problems and Achievements." *Library Management* 5, no. 1: 1–47. https://doi.org/10.1108/eb054866.

Schultz, M., and J. E. Doerr. 2011. *Rainmaking Conversations: Influence, Persuade, and Sell in Any Situation*, 2nd ed. Hoboken, New Jersey: John Wiley & Sons, 2011.

Stallings, W., and L. Brown. 2012. *Computer Security Principles and Practice*, 2nd ed. New Jersey: Pearson.

Haije, E. 2017. "Top 20 Best Project Management Systems."

Bridges, J. 2018. "The Real Value of Project Managment."

Kerzner, H. 2017. "Project Life-Cycle Phases." In *Project Management: A Systems Approach to Planning, Scheduling, and Controlling*. 12th ed, 62. Hoboken, NJ: Wiley & Sons, Inc.

Project Management Institute. 2017. *Guide to the Project Managment Body of Knowledge (PMBOK Guide)*. 6th ed. and *Agile Practice Guide (English)*.

Tucker, E. 2015. *Business Continuity from Preparedness to Recovery: A Standards-based Approach*. 8th ed. Waltham, MA: Elsevier, Inc.

Reynolds, G. W. 2014. *Ethics in Information Technology.* Massachusetts: Cengage Learning.

CYOP Consulting. n.d. "Approaches to Implementing ERP Systems."

Laudon, K. C., and J. P. Laudon. 2016. "Information Systems in Global Business Today." In *Management Information Systems,* 12–13. Massachusetts: Pearson.

Derenskaya, Y. 2018. "Project Scope Management Process." *Baltic Journal of Economics Studies* 4, no. 1: 118–125.

Leyman, P., and M. Vanhoucke. 2015. "A New Scheduling Technique for the Resource-Constrained Project Scheduling Problem with Discounted Cash Flows." *International Journal of Production Research* 53, no. 9: 2771–2786.

Warburton, R., and D. Cioffi. 2016. "Estimating a Project's Earned and Final Duration." *Internal Journal of Project Managment* 34, no. 8: 1493–1504.

Hazir, O. 2019. "A Review of Analytical Models, Approaches, and Decision Support Tools in Project Monitoring and Control." *Defense AR Journal* 26, no. 1: 84–88.

Dashkov, R. Y., and A. V. Tislenko. 2018. "Monitoring and Control Systems of the Stakeholders Activities for Project Based on the Earned Duration Management Method."

Choi, E. 2018. "Facilitation Course Connection and Transitions to Project Closure in Service Learning." *Journal of Experiential Education* 41, no. 4: 411–424.

Ruehl, C., and D. Ingenhoff. 2017. "Communication Management 2.0." *Journal of Communication Managment* 21, no. 2: 170–185.

Rozzani, N. E. 2017. "Risk Management Process." *Research in International Business and Finance*: 20–27.

Daigle, H. 2016. "Application of Critical Path Analysis for Permeability Prediction in Natural Porous Media." *Advances in Water Resources* 96: 43–54.

NC State University. 2017. *Supply Chain Resource Cooperative.*

ABOUT THE AUTHOR

Daniel grew up in the township of Gardnersville, located in West Africa, Liberia. He is the third of four siblings. He fell in love with writing at the age of fourteen due to his dad encouraging him to read every day. He currently resides in Denver, Colorado, and holds a master of information technology management (MSITM) from the Colorado State University-Global as well as Network Security (NSE) Certification from OPSWAT Academy in California.

He primarily spends the majority of his time on research in education, technology, and leadership development. Daniel also enjoys being with family, watching sports, and listening to gospel and soulful music.

Haven't spent almost ten years working in the IT industry, Daniel has firsthand experience about the leadership and communication challenges facing technical and nontechnical leaders across the industry. The historical and contemporary gap created by the leadership and communication challenge inspired him to not only write this book but to also begin an integral process of working with business and IT leaders to transform leadership and communication across the industry through a focused approach, including Managing Inherent Technical Isolation.

CPSIA information can be obtained
at www.ICGtesting.com
Printed in the USA
LVHW101654060922
727695LV00002B/369